THE MYTH ABOUT
EVOLUTION

I0616354

All inquiries should be addressed to:

Book Domain LLC.
543 E Louise Dr Phoenix, Az 85050

Ordering Information:

Amount Deals. Special rebates are accessible on the amount bought by corpora-tions, associations, and others. For points of interest, contact the distributor at the address above.

Printed in the United States of America.

ISBN-13 Paperback   978-1-970309-23-2
    eBook   978-1-970309-22-5

THE MYTH ABOUT
EVOLUTION

*"Any ordered system, including the Universe,
must have a fixed reference and
be of Intelligent Design."*

*This law is fundamental for the development of life on
our planet and any ordered system that we undertake.*

JOHN CONSTANTINE CAPLETON

BOOK DOMAIN LLC
Publish to Perfection

CONTENTS

ACKNOWLEDGEMENTS

INTELLIGENCE DEFINES ORDER
"IT IS ALL ABOUT REFERENCE"

To my wife, Cheryl, for the extra time I had, between chores, so that I was able to focus on writhing this book.

Thanks to our three daughters Lisa, Stephanie and Marissa for taking the time to review the draft, giving me feedback to help ensure it made logical sense and that the basic concept would be readily understood.

Lisa R. - For encouraging me to write the book and offering to assist me in the review of the draft.

Jenn - Thanks for your advice on rearranging the topics in order of technical complexity, making it easier for the reader to be gradually introduced to the various concepts.

Pete -You were always willing to helpme find the Bible passages referred to in the narrative to better show the obvious relationship between scripture and science.

Dan Caffese -Thanks for honing my understanding of who GOD is and how he relates to us.

Jack -Your prophecy about the message you received from the Holy Spirit and your assistance in reviewing the draft, gave me the confidence to write this book.

FOREWORD

Born in Oracabessa, a small port town on the northeast of Jamaica, just thirteen miles east of Ocho Rios, John was aware of the world beyond his door from a young age. He attended boarding school in Kingston as a boy, opening the doors for later studies in the United Kingdom, where he received his master's in engineering. Following emigration to the United States, he worked as a Risk Control Consultant until he retired in 2014.

John values working with his hands and has often been found completely absorbed in his projects. As a child, John enjoyed drawing and painting as well as designing and building models. In his teenage years, he developed an interest in electronics and moved on to making amplifiers and preamplifiers. John thrived on gaining knowledge and applying what he had learned to develop his craft. He later combined those skills and made enclosures for speakers, seeing his projects through from conceptual stage to completion.

An insatiable curiosity about how things work has encouraged an almost lifelong pursuit of the answers to what initiated the universe and the forces needed to maintain order. Adamant in his belief that all natural occurrences are governed by logic, John always knew there must be a rational explanation for existence.

Having raised three daughters, John is currently enjoying retirement with his wife and their two dogs in Arizona. His passion for creating has remained with him, and he still paints and designs & builds furniture.

MY EARLY YEARS

M Y HOMETOWN OF Oracabessa is small port town on the northeast coast of Jamaica. For those who are familiar with the popular vacation spots on the island, it is thirteen miles east of Ocho Rios.

The name of the town evolved from the original Spanish name Cabeza Ora, or Golden Head, which the Spanish named it, because of the beautiful sunsets over the bay. The sun sets on the side of the bay over the projection of the coastline. In English, the adjective comes before the noun and so the name became Ora Cabeza and spelled, 'Oracabessa'.

Jamaica was colonized by the Spanish in 1494. The British captured it in war and colonized it in 1655. However, many of the towns in Jamaica still have Spanish names. Jamaica gained its independence from the British in 1962.

Oracabessa was a very popular banana shipping port in the 1950s and 1960s. You may have heard of 'The Banana Boat Song' by Harry Belafonte. This is the operation he was singing about as bananas were checked before being loaded onto ships in the harbor for export to England and other foreign countries.

It was a popular destination for tourists visiting the island and staying at the nearby resorts. People would come from all over the world to see this operation which would go on for two to three days, every week.

It was a day to look forward to as it brought out some of the very interesting characters in the town and nearby villages to display their unique talents, some in return for money from the tourists.

One of my first memories of Oracabessa was looking at the vast ocean, from our front yard, and seeing the blinding reflection of the sun on the water, as it always did in the early afternoon.

One day, as I gazed at the water, I said to myself, "I am really alive. What I see is real". The reflection was so blinding that one could not look at it for more than an instant without being forced to look away. I believe it was from that time I developed the desire to try to understand not only the force behind the universe we live in but also the reason for life itself.

(Above was the view of the port operations from our front yard)

This was also the town that Ian Fleming chose as the site for his home in Jamaica. He wrote many of his James Bond books here.

One of his books was named after the property, Golden Eye. The property was previously named, Rock Edge, as I would hear my family refer to it, when it was owned by a friend of theirs.

My grandparents had a small property surrounded by his property and we had the privilege of using his private beach for all our summer vacations from school, along with our friends. We had a lot of fun spending time at the beach, as he was never there for more than two months of the year and never during the summer months. We took over the beach the rest of the year.

I remember my first and only encounter with Mr. Fleming. I was about 6 years old and at my grandmother's house. I saw him walking down his driveway as he sometimes would on his exercise walks and to survey his property. He was always only in a pair of shorts, sandals and smoking a cigarette.

I had seen him before and recognized him as the owner of the property surrounding my grandmother's. He always came about the same time every year. This was where he could relax and write his books without any interruption.

This time he stopped and asked me if I could call 'Mrs. Simmit'. I knew that he was referring to my grandmother, 'Mrs. Smith'. He called her Mrs. Simmit because that was how the local people pronounced the name 'Smith' and he had obviously been told her name by one of his property employees.

I had no idea why he wanted to speak with my grandmother but I went to call her and she immediately came to see what he wanted. As I stood beside her, he first asked if she was Mrs. Simmit to which she responded affirmatively.

My grandmother was a retired school teacher, well respected in the town and well spoken. She spoke perfect English and demonstrated her command of the English language in her response to him.

What he had come to complain about was that the fence at the property line, between the two properties, was broken and there was no longer a clear demarcation. I think he wanted it repaired and she agreed to have it done. However, at the end of the encounter she corrected him concerning the incorrect pronunciation of her name. She told him her name was "Smith" and not "Simmit". In addition she scolded him on his source of information, indicating that if he wanted accurate information about her and her property he should ask her and not his employees.

I was proud of how she stood up to 'Commander Fleming' as he was known by the people in the town. You see, he was an Officer in the British Navy during World War II and was well connected among the British aristocracy. Some of the people who stayed at the property were Princess Margret, Sir Winston Churchill and Sir Anthony Eden, another of Britain's Prime Ministers. When they were there, we felt safe, as a security guard was posted at the gate 24/7. We would give them snacks and meals as well as shelter on the veranda, when it rained.

That said, I must commend Mr. Fleming on his handling of an issue with the property title on his last visit to Jamaica. My grandfather had bought the property from a family friend, the same person who sold the property to Mr. Fleming. Unfortunately, there was no proper separation of the two properties and no separate title drawn up for my grandparent's property. As a result, she had no legal claim on the property and my grandfather was now dead.

She got legal council requesting that Mr. Fleming sign over her section of the property to her. He could have easily refused or fought it in court. But he did not! On his next visit to Jamaica he signed her section of the property over to her. That was very gracious of him and I have never forgotten this and respected him for that decision. That was his last visit to Jamaica as he died in England soon after.

Memories of my early life are now somewhat spotty but I do remember those incidents that made a permanent impression on me. Looking back to when I was about five years old, I was in the living room alone one day listening to the radio. I started wondering, how could someone be talking, also singing and music coming from this little box on the table. There had to be some little people and a band somewhere in there. I then got a knife from the kitchen, turned the radio around and started unscrewing the back cover. I could feel the heat being generated by the vacuum tubes, as in those days they did not yet have transistors. Vacuum tubes are not only hot but they operate with moderately high voltage. I looked inside and did not see the man or the band so I decided to try to move some of the components aside that might have been blocking my view. Somehow I touched what must have been the power supply and experienced my first electric shock. After recovering from the 'shock', I carefully replaced the back cover but told no one about what had happened.

Throughout my life I have always had a passion for designing, making and testing things to verify if my design was successful. I have built model cars, planes and boats. I liked boats because I grew up with boats in my home town, Oracabessa.

I once challenged myself to build a full sized boat from pine wood. This was one with a keel, starting with a wooden frame and adding the panels to finish it. It was 19 feet long and about 4 feet wide.

I had just finished high school but had not yet decided to go to college. I needed something to do and this seemed to be a good project. I got the raw materials from the local hardware store. Since I was not working, my mother provided the funding which she did without complaining even though she could hardly afford it. She was concerned about me and would have done anything she could to help me get through this rather difficult period of my life. She opened an account with the local hardware store and whenever I needed tools, lumber or nails I would go and get the necessary supplies. With all of this, I did not finish the boat as I left Jamaica to go to college in England before completing it. A local fisherman bought it from my mother after I had left.

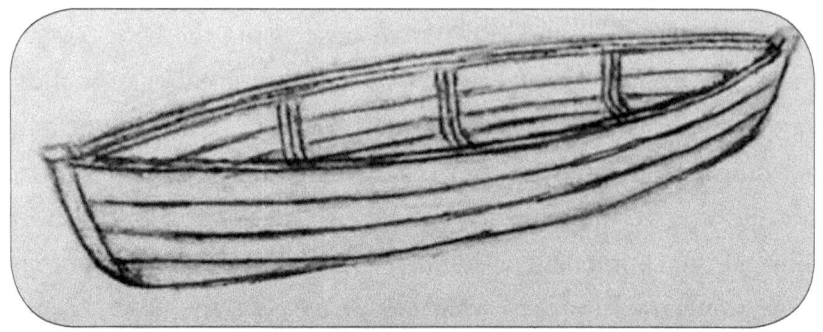

My Pine Boat

MY FIRST OIL PORTRAIT

It was at this time in my life that I also had time to practicemy oil painting. I painted a portrait of my sister from a black and white photograph that she had sent to us from London where she was attending law school. (see adjacent portrait)

I was then about 19 years old and needed to make some important decisions about my life. I was the 'black sheep' in the family and going nowhere.

It was during my last high school years that I started to explore various theories or philosophies of life including Eastern Religions and even 'The Occult'. I needed to find a philosophy that satisfied my basic understanding of life at that time.

One afternoon during the time I was experimenting with occultism, I had just come from school and started reading one of my books. I came upon a chapter that described the steps on how to teleport one's spirit. According to the book, your spirit would leave your body and go anywhere you desired. I followed the steps indicated and something did happen, but not what I expected.

I was laying on my back in bed and going through the steps when I realized that I could not move. I was fully conscious and awake and so was beginning to get really scared. No matter how I tried, I could not move.

I was totally aware of everything going on around me. I could see the curtain by the window slowly rising and receding. I could hear the wind as it slowly moved the curtain and then die into the ambient outside background noise of the birds and other random sounds. I began to panic but could not call for help as I had also lost my voice. I have no idea how long this lasted and it may have been only a few seconds, but it appeared to last forever. With adrenaline building up in me from my state of panic, I managed to move one arm, and then the other and finally my whole body.

That was the last time I dabbled in 'The Occult'. I felt so out of control and did not ever want to feel that way again.

This period of my life was very disappointing to my parents, particularly my father, who was an attorney and was of the firm belief that a good education was essential to one's success in life. He wanted me to go to college immediately after high school but I was definitely not ready.

I remember him telling me that I had no ambition and that artists never made enough money to adequately support themselves, much less a family. He would give me extra lessons in Latin as he thought a person was not fully educated if he or she was not familiar with the language from which most modern languages were derived. Classes were held on our veranda where all my friends and neighbors could see. They would make fun of me by making faces from a distance as they watched me squirm when I could not answer a question on the day's lesson.

My father was self taught after leaving high school. He taught himself law, getting all the necessary courses and exams by mail from England. He was very self motivated and expected his son to do even better than he. I was a disappointment to him for most of my teenage years but after attending engineering school and getting a postgraduate degree, I think I might have redeemed myself in his eyes. He even offered to pay for my Masters course but I had already been granted an extension of my government scholarship and did not need his help anymore.

Some time before he died, he told me that he thought engineering was such an interesting occupation and he always wondered how a massive, steel oil tanker floated on water. I was glad for the opportunity to explain 'Archimedes Principle' of flotation to him. I don't think he really understood, although he believed what I told him.

(Archimedes principle states that an object will float on water when it displaces the volume of water equal to its weight. Water weighs 62 pounds per cubic foot and so it would support an object weighing 62 pounds if that object displaced one cubic foot of water when placed on its surface. The oil tanker, therefore, displaces the volume of water equal to its weight. This principle is true for any liquid)

I remember my mother being a very nice human being. She would give anything she had to those in need and was well liked in the town. I needed someone like her to keep me balanced. I was very close to her and felt as if I could talk with her about anything and she would not judge me.

She was the secretary at the United Fruit Company, one of the banana shipping operations, the main industry in the town. Her job included administrative work such as typing documents and making out checks to pay the banana suppliers who sold their produce that was harvested weekly.

I would sometimes watch her as she typed checks for some of the suppliers and could not help noticing the amounts on some of the checks. These people were making tens of thousands of 'British' pounds per week and in some cases hundreds of thousands. This was a lot of money in the 1950's.

Whenever I think about the wonders of word processing, I remember my mother in those days of carbon paper. Typed copies were made using carbon paper. If you made one mistake you had to stop and erase the mistake on the original and all the copies, of which there were sometimes 5 or 6. It was not only time consuming but it was also messy. Although she made few mistakes and was good at making the corrections, I could still see the look of frustration on her face when she made a mistake, as she knew what she now had to do.

One plantation owner, for whom she would sometimes do special secretarial work, had one of the larger properties with hundreds of acres of bananas and so he got a healthy weekly check. On 'Banana Days', I would see him come into the office, from his chauffeur driven Mercedes Benz, to see how things were going and to talk with the office staff. He would then go out on the landing in front of the office overlooking the port to watch the shipping operations. He also owned two launches powered by Mercedes Benz engines, used for fishing and for pulling the banana boats out to the ships. A Mercedes Benz mechanic would visit annually to perform routine maintenance on the boats.

It was on one of these launches that we were promised a Saturday afternoon ride for the office staff and their families. I was then about 8 years old. I had looked forward to this day. It was a beautiful afternoon with the sun shining brightly on the ocean. We waited and waited but he never came. There must have been a good reason why he was not able to come but since there was no telephone at home at the time, we did not know what had happened. This was probably one of the most disappointing experiences of my childhood and I still think about it. I believe he was really sorry for disappointing us as he would give my mother the annual calendars and catalogues from The Mercedes Benz Company to pass on to me. It worked because I always looked forward to getting my annual catalogs and calendars. My mother always gave me the United Fruit Company magazines with pictures of the fleet of all their ships. That was a real treat for a young boy.

I thought Mercedes were the most classy cars. One day a traveling salesman came into town and stopped at the office. Can you guess what he was selling? Battery operated toys which were the latest upgrade from push and wind up toys. He had fire trucks and cars but there was this red 'Gullwing' Mercedes Benz sports car that really caught my eye. It was not cheap but somehow I persuaded my mother to get it for me.

After she bought it for me I was completely preoccupied with this toy. I played with it from the time I got it until I met with my friends that afternoon. I was looking forward to showing off my new car. They passed it around so each could examine it and play with it. Then it came to one boy's turn.

You know, in every group there is always one that you wonder whether or not he is really a genuine friend. It was now his turn to play with my car. He picked it up, looked at it and then dropped it.

I did not know whether or not it was on purpose but the car stopped working.

I went home with my broken car and took it apart in an effort to fix it but could not get it to work. I then took it to my friend, Maxie, to see if he could fix it for me. Maxie was like a big brother to me and my mother liked and trusted him. All I had to do was tell her that I was going over to Maxie's and she would ask no further questions.

We worked on the car for several hours but still could not get it to work. We tried everything but nothing worked. Eventually, we had to give up but I have never forgotten the disappointment and the regret for having shown my car to that friend.

I believe this was when the seed was sown for my interest in engineering. I made every effort, at every opportunity to learn about how things work and how to design and build things.

In the early years my models were built from wood, metal, and plastic. Later, I used fiberglass and carbon fiber. These are typical manufacturing materials and so I would experience the same challenges working with these materials as one would in a manufacturing facility.

My Education

The system of education in Jamaica started with kindergarten, then elementary school at about the age of six years. We were taught the basics-English language, English literature, Arithmetic, Geometry, Algebra and some basic Science. In addition, there was Home Economics for girls and Woodworking for boys.

At the age of about ten years, you were now prepared to take The Scholarship Examination which one had to pass in order to be accepted into High School.

We had very good teachers who would spend extra time with students after school to help them prepare for the scholarship exams. I was a very dedicated student and remember doing 100 math problems daily, under test conditions, in preparation for the scholarship exams.

I had now taken the scholarship examination and was awaiting the results. The morning when the scholarship results were published in the Jamaican news paper, The Daily Gleaner, when I scanned the list, I did not see my name. I thought I had done well enough to pass so I was extremely disappointed. I was very worried as I knew that my parents, especially my father, would not be happy and I would have a lot of explaining to do. It wasn't too long before one of my friends who was also scanning the news paper for his name said 'Si yu name ya'. This is Jamaican dialect for 'Your name is right here'. I was so relieved! The reason I could not find my name was that I was looking for it in the wrong place. I had been awarded a special scholarship and my name was printed in a separate place with some other students who had also been given special recognition.

Everyone now had great expectations of me in high school but I performed as an average student and definitely not up to the level of expectation of my father. After the first two years at the first high school, he transferred me to a boarding school in Kingston, as he thought this would be a better work environment for me and a place where I could focus on my studies.

I hated boarding school and continued to be an average student but did pass my Ordinary and Advanced level examinations in preparation for college. The last two years of high school, I stayed with an aunt and uncle who lived in Kingston. I would no longer be attending boarding school.

After leaving high school, for the first year, I stayed at home in Oracabessa and explored my hobbies including oil painting, wood-working and electronics. Electronics also became one of my interests as I have always liked music and would listen to the radio to the popular music of the time. I built amplifiers and preamplifiers, from circuits published in Popular Electronics magazines, to enhance the fidelity of the music.

Moving forward to when I was now about 20 years old, I got a job as a teller in a bank. This was probably not the best choice for me as in those days the bank employees were known for their drinking and partying and I was now one of them.

I worked at the bank for less than two years, then got another job as an auditor for the Courts Office. This lasted only a few months as I soon realized that I needed a change and applied for and got a government scholarship to go to college in England to do engineering.

I left Jamaica for England in 1970 to continue my higher education. I spent my internship with a company named Metro-Cammell. Here, they built buses and trains including those used in the London Underground Transport System (Tube). They also built buses and trains for the Commonwealth countries, including Jamaica and India. My training alternated between six months there and six months in college, for my first degree.

At Metro-Cammell, I was trained in each department, initially at the Training Center where I learned to operate production machines such as the lathe, milling machine, shaper and surface grinder.

Before I was allowed near a machine, I had to learn to use my hands, with great precision, filing a piece of steel until it was perfectly square in all dimensions. This was a small piece of mild steel, commonly used in manufacturing, rough cut from a plate about

1/8" thick, 6" long and 2" wide. I was given a hand file to transform this rough piece of steel to a tolerance of 5/1000" on all dimensions. It seemed to me like an impossible task but there were other trainees who went before me and had their finished work displayed as proof that it could be done.

You may be wondering how was I to know when I had achieved success? In addition to the steel and the file, I was given a reference, called a square, Fig R1a. This is a reference tool that is perfectly square and perfectly smooth and flat on the reference surfaces. It is 'L' shaped. The perpendicular section is at a perfect 90 degree angle. This was my reference for squareness and flatness. This is one of the reference tools that is still invaluable even today. To measure the dimensions, a vernier caliper was used. This is a highly accurate measuring tool with an accuracy to less than 1/1000th of an inch.

To use the square as a reference, when you held it up to the light against the work piece, if you saw any light coming through between the contacting surfaces, then you had to continue filing because the side you were working on was either not flat or not square. To achieve success, each of the four corners had to be square relative to the adjacent corners; as well as the thickness, relative to the larger flat surfaces; See Figs. R1a and R1b through R3a and R3b below.

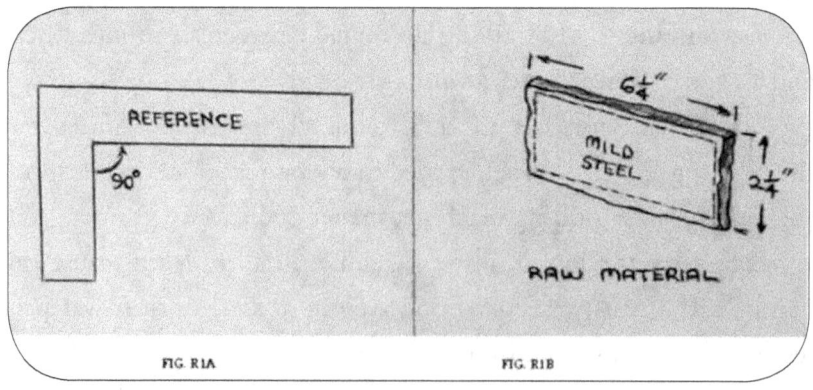

FIG. R1A   FIG. R1B

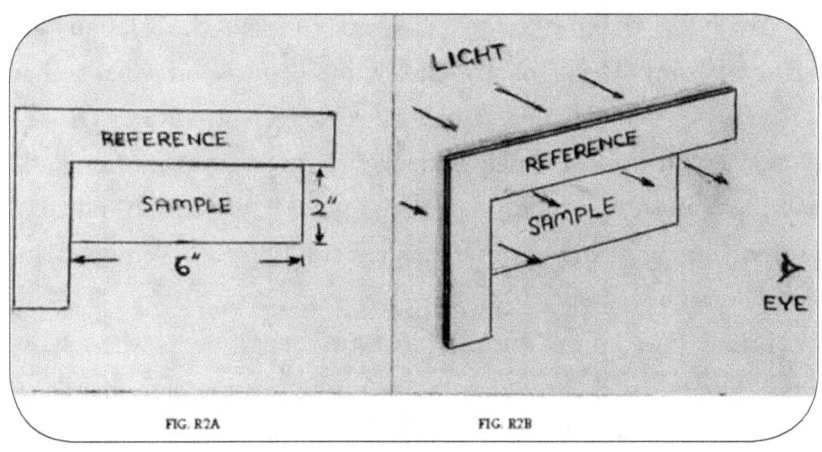

FIG. R2A   FIG. R2B

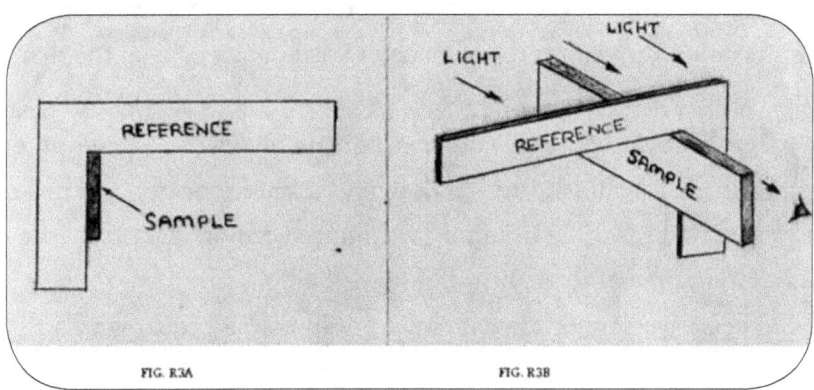

FIG. R3A   FIG. R3B

I think it took me two to three weeks to achieve success but, once completed, it did give me a sense of accomplishment.

I did not spend all of my time doing this project as that would have been too much. My muscles would get fatigued holding the small file, trying to accomplish the task. To break the monotony, I was given other projects to do but was expected to eventually complete this project.

I could not have achieved success without a reference with known accuracy. I had to complete this project before I was allowed to learn how to operate the machines, so it was important.

After the training center, I spent time in the Design and Manufacturing Departments and wrote reports on what I had learned in each of the respective departments. Research and Development worked closely with Design and Manufacturing and I had the opportunity to test new designs and verify tensile and compression strengths under different load conditions. I even did some simple designs myself.

I then moved on to the administrative departments where I also wrote reports on their operations. Training was therefore hands on and the best way to become familiar with the design, manufacturing, and administrative processes.

The hand filing exercise taught me to develop patience and an appreciation for what it took to achieve accurate results in the manufacturing process.

My Masters thesis in engineering was to develop a computer program that would aid in the design of a stable control system for a production machine in order to produce accurate parts. The title was 'Computer Aided Control Systems Design'.

The active computer program would be designed to send a signal to a proportional control servomechanism that would control the tool during the machining process. The output would result in the tool accurately copying the input signal. For testing, the output was connected to an oscilloscope which displayed the output in analogue form.

I developed the program and tested its stability by inputting test signals such as a sine wave, step response, ramp and other waveforms.

The first test of the program resulted in failure. We later found that all calculations in the program were correct except for the scale on the visual display, on the oscilloscope, which initially showed a flat line with every input signal. I controlled each input signal and knew the expected output, but the output did not match the input.

Then, my professor got the brilliant idea to change the scale on the 'Y' axis. This made all the difference because the proper graph now appeared on the screen.

This indicated to me that every step in the journey is critical. The entire sequence has to be accurate and complete before the final goal, and hence success, could be achieved. There has to be a set sequence of events starting from a fixed reference with all of the steps accurately completed before success can finally be achieved. This is not a random process.

Success does not come randomly. It takes a determined, focused effort. You have to know the desired outcome and take the sequential steps and make the necessary effort to achieve it. Time is not the most important element even though time is a necessary component. It became apparent towards the end that even if all other steps had been correct it was just as important for the final step also to be correct.

Developing this program made me realize that the slightest error could result in failure. Every step had to be correct for the whole system to be successful.

After completing my education in England in 1975, I went back to Jamaica and worked in the manufacturing industry for about two years before immigrating to the United States, in 1977.

I got married in England before returning to Jamaica. Our first daughter, Lisa, was born there, less than a year before I graduated from The University of Birmingham. After returning to Jamaica, our second daughter Stephanie was born. It was fifteen years later that our last daughter, Marissa, was born in Glendale, Arizona where we currently live.

I had never been truly religious but always believed there was a unifying force in the universe. This force existed but, to me, it was always far away, somewhere up there, out of reach, in the heavens.

When I was a young boy, I went to church on Sundays because my parents required that I learn about GOD. To them, it was the right thing to do. When I was on my own for the next thirty years, I only attended church at Christmas and occasions such as weddings and funerals.

I still believed that there was a unifying force but it was still out of my reach. To that force, I was insignificant and unimportant or possibly, nonexistent. I only thought of GOD occasionally when I was in a reflective mood.

I was over fifty years old when things dramatically changed in my life. A series of revelations concerning life and the process of creation made everything begin to become clear and make rational sense.

When I look back at my early life, I can see how it has influenced my entire life. This included my choice in career and my chosen hobbies. It is as if these were chosen for me, and all I did was to fulfill the purpose. This purpose was not of me, so it must have been from a source outside of me; a higher source, the source we refer to as God.

INTRODUCTION

WE ARE ALL faced with the same questions. How did this all begin? What is the purpose of our existence? What part do we play?

We are born into a world that is initially foreign to us. We have to learn everything about this material and spiritual world. We all start at the same point, birth, and then gradually grow through childhood, adulthood, and finally, old age, if we are blessed with an average lifespan.

Some believe that this is a random existence with no real purpose. Others believe there is a real purpose for our being here and our lives are a fulfillment of this grand purpose.

In this book, I will look at both of these beliefs in an attempt to determine which best fits how our world manifests itself and the possible source of its existence. What we will come to find is that there is nothing random about the order that we see and experience. Order and randomness are mutually exclusive. The laws that control our material world are fixed and never change. Together, these generate universal order.

As we discuss these laws, we will see that they work together toward one purpose: to maintain harmony in the universe. This is an indication that the source of these laws is also fixed and never changing. This is the only conclusion as only such a source could generate and maintain such laws. We will determine that only God has the attributes to be the source of our existence.

To help guide us through this life and help us determine our purpose here, we should take into consideration where we were born and the gifts or talents given to us, as well as our experiences from childhood to adulthood and old age. These shape our lives and teach us about the world. I will first share with you some of my experiences, from birth through adulthood and retirement, as examples of how these experiences shaped and directed me toward my purpose in this life.

THE ABSOLUTE AND FUNDAMENTAL LAW OF ORDER AND TRUTH

"Any ordered system, including the Universe., must have a fixed reference and be of Intelligent Design"

THIS LAW DESIGNED the universe to be an ordered system. Without this law there can be no order. If this law is broken, disorder will be introduced into the system and there will now only be partial order. In addition, once disorder is introduced, order will be affected and will start and continue to erode. This is the Universe in which we now live.

As indicated above, when we look at our universe, we see both order and disorder.

Let us first examine order. If we look at any ordered system, we see that there are certain attributes that we recognize for us to conclude that the system is ordered. Some of these are listed below:

- Symmetry
- Harmony
- Consistency
- Theme

- Balance
- Stability
- Fixed Cycles
- Fixed Sequencing
- Aesthetics
- Structure
- Predictability
- Pattern
- The presence of purpose
- The presence of limits
- The ability to be defined

An ordered system may be recognized by having one or more of the above characteristics.

We see these characteristics all around us. They are an indication of how order manifests itself.

When we try to develop an ordered system for ourselves, there are certain things that have to be taken into consideration. We first develop a plan with a fixed reference, start, and then build the system until it is complete. Depending on the system we are creating, there are specific sequential steps that must be taken. We also have to recognize when the system is complete and functioning as designed. This is not a random or chance process but takes intelligence. It is our intelligence that defines the system. There are no exceptions to this process.

The basic requirements for the development of an ordered system are a fixed reference, and intelligence to define sequencing. Intelligence fixes the reference and defines the sequential steps needed to develop the system. This is true for any ordered system made by man or other intelligent living creature.

To accomplish this, the intelligence must be external to the system so that it has an objective perspective. Otherwise, it will be unable to develop an independent, definable system. If we are a part of a system, we are influenced by that system and thus unable to look at it objectively. Before an ordered system is initiated, it must be planned. Planning can only be initiated from outside the system.

If we now look at disorder or randomness, we find that neither can be accurately defined. If something cannot be defined, it cannot be reproduced. At the nano level (fundamental level), atoms and molecules can be accurately defined and so are ordered systems. This means that the building blocks of all matter are ordered. From this we can conclude that they are of intelligent design. All ordered systems must, therefore, incorporate fundamentally ordered matter and components so that the complete system may be ordered.

By the same reasoning, the universe must be of intelligent design because it is an ordered system. Intelligence is that which defines order. As far as matter is concerned, both disordered and random matter are the raw materials for creating an ordered system. It takes intelligence to use these raw materials and transform them into independent ordered systems. It also took intelligence to create the fundamental components as ordered systems (atoms and molecules).

We can confirm this truth by examining any ordered system. Order cannot be randomly achieved. Neither can it be achieved by chance. It takes intentional, determined acts to achieve order. Additionally, only the intelligent designer is able to recognize when the system is complete and functioning as designed.

If we examine spiritual order, the principle is the same. This means that the statement below is 'Absolute Truth'.

'Any ordered system, including the universe, must have a fixed reference and be of intelligent design.'

Now, let us look at The God of the Bible. God has the following attributes as well as many others.

- He is Eternal
- He is the only God and there is no other
- He does not change

These attributes would have been fundamental in initiating an absolute, ordered system.

1. In order to initiate an absolute ordered system, there has to be an Eternal God. He was not created. He always existed in the form of infinite energy and intelligence. We cannot understand the concept of eternity because we are trapped in a temporal system.

2. He must be the only God as there can be only one ultimate source of control in any ordered system, especially if there are attached subsystems. All the systems must operate in harmony and so must be under a singular control. Otherwise, there will be chaos and disorder.

3. God does not change. This is critical for the development of any ordered system, again, especially for the development of the most complex systems such as Heaven and the Universe. If the reference changes in an existing system, disorder develops. In the material world, a reference change manifests itself as material disorder. In the spiritual world, it manifests itself as Sin.

Order is the product of intelligence. This is how we identify an intelligent designer; by the order inherent in the product's design. Order is how intelligence manifests itself. There can be no order without intelligence. Intelligence defines order. This is Absolute Truth.

Since the fundamental elements of the universe are ordered, there must have been an intelligent designer. This intelligent designer is whom we refer to as God. There can be no other.

If there were another initiating designer, the fundamental elements would not be of similar properties and structure. There could not be harmony as there would be fundamental differences preventing harmony. In the spiritual world, there could be no peace or joy since there would be constant interference from non-harmonious forces, possibly as powerful as the order we know. The peace and harmony we have come to know is from a singular source, God.

If you cannot define something, you cannot reproduce it. If you cannot reproduce it, order and hence life as we know it would be impossible. Only intelligence can define, and so only intelligence can create order and hence, life. This is not a random process. It is fundamental that a reference be set for any ordered system to be initiated. This reference cannot change. God does not change and so He is the perfect reference for the creation of order and life. There is only one Truth as there is only one spiritual reference and that reference does not change.

If you are in a material or spiritually ordered system and you lose your reference, you become lost. Once lost, one cannot recognize or locate the true reference from within the system. This is because you now have no true reference that you can use to navigate. Only some source, outside the system, who knows the true reference can direct you back to truth. This is why an ordered system can only be initiated by an intelligent source outside the system, using the reference that has not been contaminated with disorder (the True reference). The reference must be set and monitored from outside a system for it to stay true.

Perfect order is eternal. An ordered system only deteriorates if disorder is introduced into that system. Intelligence recognizes the difference between order and disorder and can choose either.

The entity with Absolute Intelligence has Absolute Knowledge and Control over all systems. We refer to this entity as God.

- You may ask, what is an ordered system. Here are some examples:
- All living things
- All things man-made
- The things made by intelligent creatures
- An atom
- A molecule
- The Universe
- Anything of which we can make sense
- Anything that can be defined
- Anything that is or can be reproduced

What does The Eternal God say about Himself:

"I am God and there is no other"

"I am the same yesterday, today and forever"

"I am eternal"

The above are critical attributes for a creator. Only God defines spiritual and ethical order. Man is given authority to define material ordered systems.

Based on the above reasoning, we can clearly see that God has revealed to us the necessary clues for us to conclude that He is the One True God.

THE MYTH ABOUT EVOLUTION

T HE FINGERPRINT OF GOD is evident in the fabric of the universe.
Here, I have discussed scientific evidence that shows, beyond a
reasonable doubt, the existence of GOD.

Introduction

This book presents a 'singular' concept that is the fabric of the universe. 'Singular' meaning its origin is a single source. It manifests
itself as a recurring theme throughout the universe as well as in our
daily lives. It is so simple that we take it for granted, yet it represents
a connection between matter and life. Both exhibit the same basic
'TRUTH' as to their origin, a truth that is reflected in everything we
do, in the way we think, in fact in everything that may be defined
as being ordered.

When you do an investigation of an incident such a robbery,
you may not initially know who committed the crime. You must
therefore start with the crime scene to give you clues as to what
happened. The things you look for are fingerprints and now, DNA,
some characteristics that are unique to a single individual. The
'MO' - modus operandi- mode of operation, is another of the ways
of identifying the perpetrator. This shows the pattern in which different individuals do things in their unique styles. Also, you look for

articles that can be traced back to a single source. If you can connect all the clues to a single person, then you will have found the perpetrator. We will now use this same reasoning to try to determine the origin of the universe.

All of nature follows fixed laws, the characteristics of which we have determined by careful observation and experimentation. We cannot change these laws so we have learned how to use them to our advantage by discovering ways to compensate for their effects, to achieve a specific outcome or to enhance an outcome.

Only ordered systems with fixed references can be used to describe or simulate the universe in which we live. This is an indication that the universe itself is ordered and has a fixed reference. By observation, we know that the universe is an ordered system. It is also evident that all references are connected to a singular source. In other words, we are all connected. This is evident in that we are fully aware of each other and can interact with each other and our environment.

Nature's laws include those relating to gravity and nuclear forces, thermodynamics and electromagnetism. We have even developed equations to define them mathematically and to determine the mathematical relationship between them, making it easier for us to use them in practical applications. These applications include research in physics, chemistry, biology and nuclear science and have resulted in all scientific inventions.

I think everyone should try to understand this concept. While reading this book, use your own reasoning, experience and judgment to make a determination as to whether you believe this to be 'TRUTH'.

If you are a believer (in GOD and JESUS CHRIST), this book should give you more confidence when sharing your convictions with others. If you are not, it should raise questions about evolution, if that is what you currently believe. If you are an atheist, it should give you some food for thought about the existence of the universe and how it came into being. If none of this matters to you, it should still be an interesting read.

Here, I have looked at a typical human lifetime and examined the human experience from birth. We all have the same experience of becoming aware of our existence and preparing ourselves for a lifetime in this world.

We are born with the tools necessary to interact, appreciate and learn about the world in which we live. We have a brain that makes us intelligent beings and five senses that allow us to interact with our environment. Our five senses provide information that is fed to the brain for interpretation. They are all monitored and controlled by one central controller, the brain. This was all given to us without any contribution on our part.

Have you ever wondered where you came from and the purpose of your being on this earth? I have thought about this all my life but without much success, until now. At this time in my life, I looked back to see if I could find a pattern or some indication as to my origin and purpose. I examined my life from my first memory through childhood, school, work and now retirement.

What were the most important experiences that shaped my life and made me into the individual that I am today? I believe everyone should ask themselves these questions. It would help you make informed decisions as to the direction you want to pursue in life.

At the end of your time here, will you be satisfied with the life you have lived? Is there anything you wanted to accomplish but did not? Do you still want to accomplish more?

After reading this book, you should have a better understanding and a rational interpretation of life's origin and the part each of us plays.

The Universe

Let us look at the universe and its infinite size as we observe it from planet Earth.

When you first looked up at the skies on a clear night and saw all the stars tracking in unison across the dark background, do you remember what thought came to mind? Mine was, "How insignificant I am in all this infinite expanse"! A time lapse gaze reveals a slow, choreographed movement with all the planets and stars moving in unison. Since it is the nearest, the moon appears to move against this backdrop, so massive, yet so silent.

Another significant feature of these objects is that they are all spherical in shape. This again manifests order and a common reference for all of them.

I now know that there were billions of stars in my gaze and understand the awesomeness of what I was viewing.

It was at this early age that I began to question, 'what is this all about'? One thing I did conclude is that it all manifested 'ORDER'. There is no chaos in all this giant expanse of large moving objects. It is not random because they follow cycles that have been repeated over billions of years and still continue to do so. At that time in my life I was only about 7 years old.

With this in mind, I think we can all agree that the planets and stars move in an ordered manner in space. Ordered, as opposed to random or disordered.

Now, let us look at planet Earth! As far as we know, it is the only planet, with these characteristics, in our galaxy or any galaxy we have been able to study. It has an abundance of plant and animal life that perfectly complement each other. We have a sun that makes life possible, both animal and plant. In fact, this is the only planet that we know of where life exists. This is an indication of how unique it is to have this combination existing in only one place, planet Earth.

Earth is in an orbit around the sun in a way that makes it ideal for the development and growth of life on our planet. It is as if its orbit was predetermined.

As I previously mentioned, the basic shape of a planet or star is spherical. If we examine its form, any given sphere has a fixed radius. The radius of a sphere is constant from the center, to any point on its surface. This is the simplest representation describing a three dimensional object and is, by far, the most common manifestation. The fixed reference is the center of the sphere and any straight line equal to the radius, from the center of the sphere, describes a point on its surface. It is also interesting to note that protons neutrons and electrons are described by the same formula, but at the nano level.

Life

Matter, as we know it in this universe, obeys the law of entropy, the Second Law of Thermodynamics. Entropy basically means that, if left alone, over time, matter will continue to deteriorate or lose energy. Things only improve if a source, with a positive influence, intervenes. This implies an ordered (nonrandom) source intervening. Entropy, on the other hand, tends toward randomness.

Because of entropy and decay, any man-made ordered system begins to deteriorate immediately after completion. Initially, the change is so gradual that we may not even notice it. However, it becomes noticeable over time. This change is the process of oxidation or decay and loss of energy or entropy. It will continue until a stable lower energy state is reached where the process becomes truly random. This is why we always need to maintain man-made ordered systems in order to extend their useful life.

'ORDER NEEDS CONSTANT MAINTENANCE'

L IFE IS THE only self-generating, self-preserving process that defies entropy. It takes disordered or random matter and converts it into ordered systems. Life is able to reproduce itself and has continued to do so over millions of years. This is a unique ability that is not consistent with the law of entropy that governs matter. Some even argue that life is not of this earth. One thing I think we all can agree on is that anything that has life is an ordered system.

My belief is that 'Any ordered system, including the universe, must have a fixed reference and be of intelligent design'. I challenge you to find an ordered system that does not have these essential components.

We have discovered that all living things, whether plant or animal, have a fixed design reference. This is the DNA of each species. Now, all we have to show is that an intelligent source created life.

Since we were not there to witness creation, we all have our own beliefs as to how the universe came into existence. We can, however, look at creation for evidence as to its origin since the 'fingerprint' or evidence of the characteristics of the creator is always present in any creation.

The first step is to look at all ordered systems to determine if we can find any that is not of an intelligent source. If we cannot find-any such system, then we can reasonably conclude that the previous statement is true, that order is created by intelligence and therefore life is of intelligent design.

If we look at any ordered system that has been developed by human beings, we can conclude that, without exception, it was created by an intelligent designer. If we examine any other ordered system on our planet, we see that it was created by some intelligent source. A bee hive (honeycomb), a bird's nest and an ant hill. The organisms that make these all exhibit some form of intelligence.

Now, we have looked at the universe and concluded that it is ordered. Then, by this definition, it must have a fixed reference and be of intelligent design.

Physicists have shown us that our universe originated at a single, fixed point in time that they refer to as the 'Big Bang'. Since the Big Bang (the fixed reference) developed an ordered system, then it must have been the product of an intelligent source. Only intelligence can define and create order. The latter cannot be proven but can be shown beyond a reasonable doubt. Then, if this is true, there must be a GOD, the intelligent source.

Human Life

Human life begins with two living cells fusing, rapidly multiplying and growing. In a few weeks, the cells begin to specialize to form our various organs, our body frame, the shape of each bone, and everything else that makes us unique individuals. We have discovered that this process is at the instruction of our DNA, the blueprint of life. Genes further fine tune the instruction to include family similarities that are passed down from parent to child.

DNA is an intelligent structure of cells that even has the information on how to form and develop a brain which will perform arguably the most important function in our bodies. DNA is found in each cell of your body and helps control and coordinate its development and function. Since everyone's DNA is different, it can be used to identify each individual.

In the human body, throughout life, the DNA (and genes) continues to control cell generation and development. For each individual, all cells generated are uniquely yours. This is why your body operates as a unique unit. Each cell can be identified as yours and will reject any other.

For the body to operate as a unit (in harmony), there must be one control center, the brain. Through the central nervous system, it is connected to the organs and all other parts of the body. It is the fixed reference that the body needs to function normally and as one.

It cannot be overstated how finely tuned our bodies are to the extent that even a small variation from the norm upsets so many other functions within the chain of control. When the body operates normally, it is no less than a miracle. It is designed to self maintain, an ability that only makes sense if the designer saw the big picture, including possible environmental changes that could affect its normal function.

Before memory begins, we have developed no conscious reference and so it appears to us that life only started when conscious memory was 'switched on'. Our history becomes a series of events or memories starting from the first. The first memory is fixed and has to remain so in order for us to develop a knowledge sequence and always be aware of who we are relative to that first memory. This is our experience, which defines us. This includes our family, friends, likes and dislikes, and knowledge gained both at the conscious and subconscious levels. It is what makes us unique.

To help us navigate life, we are each given aptitudes in specific knowledge acquisition in which we excel without much effort. We sometimes use these as a guide in selecting our careers in life. These are referred to as 'gifts'. They may be a hobbies or personal interests but these all make us unique in a world with seven billion people. Each of us is one of a kind.

To define how we see ourselves, we use the reference fixed by our first memory followed by other memories in sequence (knowledge). This ability is also programmed into our offspring. This is only possible because we are intelligent beings. Intelligence gives us this ability.

'We Are' - Consciousness

At some stage in this development we realize that we exist. 'WE ARE'. This is at an early age, probably between two and four years old. This is when conscious memory begins. Before this age, we are consciously and subconsciously gathering information and gaining knowledge that will prepare us for our lives ahead. Even in the womb, as we are developing, we are gaining knowledge. This is the first indication that we are intelligent beings but still unaware that we exist.

In order to recognize our existence, we have to develop a fixed reference in time from which we start a sequence of memory events associated with the order of which we are now a part.

Then, it is as if a switch trips and we become conscious. We are now aware of our existence because we have a recorded reference on which we build with each experience. All of this is programmed or automatic. Up until this point, we have contributed nothing to the developmental process. The world is already ordered, a fact we will soon discover.

Plant Life

The entire cycle of the life of a plant involves growing from a seed to a mature plant, then producing fruit and more seeds that will eventually grow into other plants to keep the cycle going. Within this cycle they provide food for animals as well as other plants.

If we look at a plant, as it blossoms to bear fruit, there are usually numerous blossoms. Bees are attracted to cross-pollinate and fertilize them. Then there is a selection process. The plant selects the most healthy blossoms for continued development into fruit. It is clear that there must be a standard or threshold by which this is determined. This is preprogrammed and the plant knows which blossoms to reject based on the probability of survival through the developmental process. This is not a random process and indicates some form of intelligent design. If a fruit gets prematurely damaged, somehow the plant recognizes the problem and rejects it, if bad enough. Again, this indicates feedback and monitoring of the process to ensure healthy fruit. This is a programmed response which is so much more than a random process.

Earth

Earth is a living planet. If we look at the analogy between human and plant life, we see that water is the common element that supports and maintains life on earth. Just like blood and sap in animal and plant life, respectively, water continuously circulates on earth giving life to its inhabitants.

Rain and snow fall onto the earth's surface dissolving the nutrients in the soil so they can be absorbed by the plants. For better distribution of water, rivers and streams flow like arteries and veins supplying life giving water to all parts of the earth. Water evaporates from the surface of the earth and the ocean to form clouds which then precipitate as rain. The cycle then continues.

Keeping in mind that animals and plants consist mostly of water, when we look at the earth, most of its surface is covered with water. There is a consistency in the design of life and the life sustaining process which appears to indicate a singular source.

Anything that is not connected to the life giving source dies. We have to be in the cycle of renewal to stay alive. This is essential in all forms of life; animal, plant and our planet, Earth.

All this relates to life in the material world, but the same is true for the spiritual world. To gain eternal life we have to be connected to JESUS, the eternal life giving source.

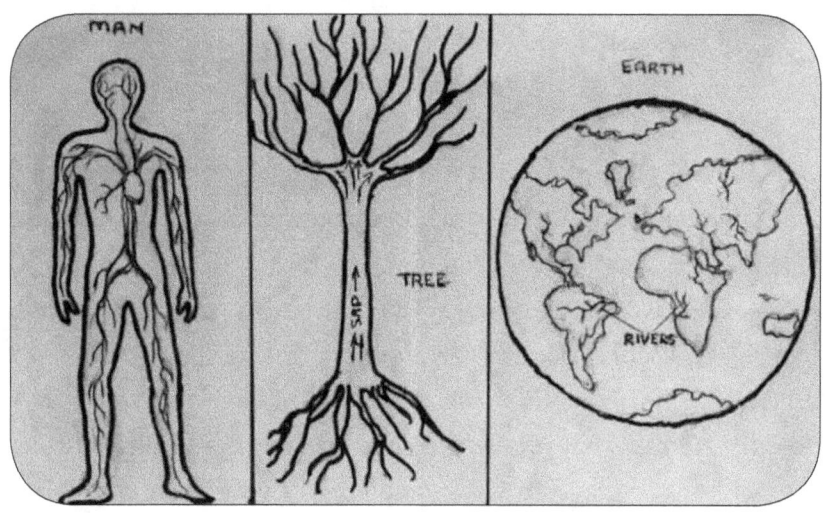

The diagrams demonstrate the similarity between the life sustaining source in the different forms of life, material and spiritual.

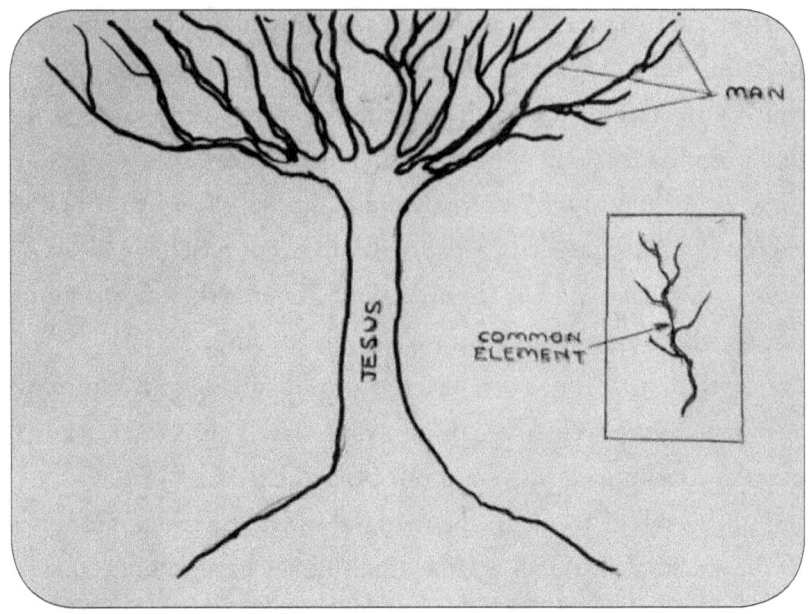

Nature's Cycles

Day and night, seasons and years are all examples of cycles. For every cycle, there must be a reference and this reference must not change otherwise behavior will be random and erratic.

We use day and night, weeks, months and years as references in time. We can only use them because they are constant and we can depend on them not to change. Wouldn't it be logical to assume that they all must have fixed references in order to remain constant over millions of years? By observation, we can conclude that this is the case.

Mathematics is used to simulate natural occurrences or events. We use this system to develop equations that closely simulate how nature operates within the confines of its laws. Mathematics uses waveforms to represent cycles.

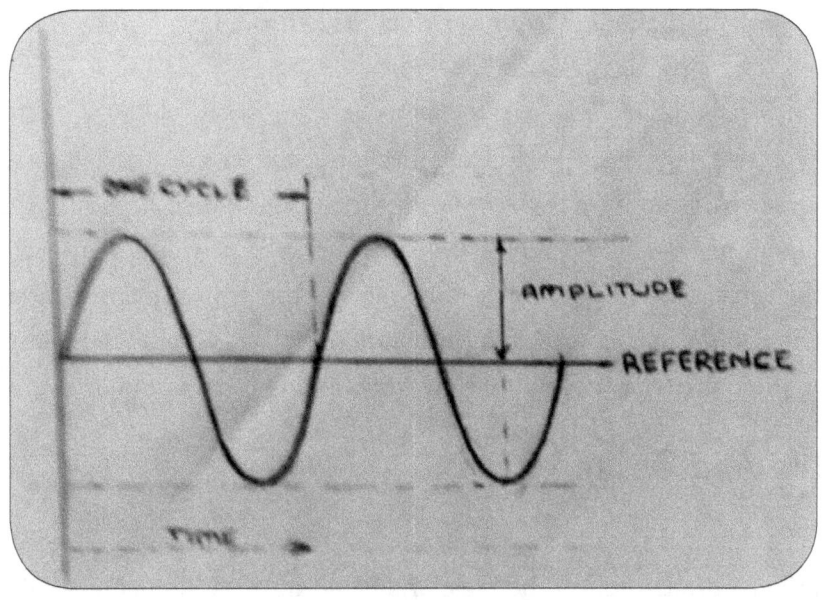

Typical Mathematical Waveform

Other examples of cycles include life and death and our daily activities. This even includes meal times. For a cycle to continue without change, it must have a fixed reference.

In the case of our solar system the sun is the reference. It influences all the planets that are in orbit around it as these are maintained in fixed orbits.

If we look at the big picture, we see that the universe is expanding. If there is a reference, there may be a point when it reaches maximum expansion and then begins to contract on itself.

But will it? If it was created from an infinitely intelligent mind there will probably be an intervention at some point. We have been told (in the BIBLE) that, this universe as we know it, was not designed to last forever.

Bible quote (The universe will be destroyed by GOD
- Rev. 21:5, 2 Peter 3: 9-13)

At the nano-level, electrons orbit the nucleus in fixed orbits. Other particles have been found in the nucleus which have specific functions controlling how the system works as a whole. However, the three most important components are the proton, neuron, and electron. Matter is consistent in nature in that it is made up mainly of these three particles. These are the building blocks of matter.

Life And Death

Life manifests itself in the cycle of life and death. Life first gives life to another living system after which it dies. The cycle then continues with the new life giving life to another, through reproduction, and then, itself, dying. Something must die in order to give life to another. This is how nature works.

Life is ordered. Order is created from randomness. Randomness is order's raw material. Death creates randomness. In order to recreate or sustain life, there must be death. The food we eat demonstrates this 'TRUTH'.

(This is probably why Jesus had to die for our sins to give us eternal life. Jesus is eternal. Only a GOD can give life eternal and so a GOD had to die to give us eternal life. Only such a sacrifice could satisfy GOD'S JUSTICE)

Bible quote - (1 Peter 2:24 ESV)

Anatomy And Physiology

The human body is the most complex living system know to man. It is a complete system with multiple subsystems that function as one. In order to function as one unit, each subsystem is connected to the central nervous system controlled by the brain.

Our body was perfectly designed. If we were to test a human body that was functioning to design specification, all systems operating within the design tolerances, it would operate perfectly.

Our body incorporates electrical, chemical, mechanical and mental systems all coordinated from a single point, the brain. Each system or subsystem is unique to an individual's body and does not function normally in anyone else's body. This is because the systems, in an individual human body, are custom designed for that particular body. The blueprint is the DNA.

A baby developing in a mother's womb has a separate blood circulatory system from that of the mother. This ensures that both human systems, although joined, operate independently. One reason being that they may have different blood types and there can be no cross contamination. In the same way, organ transplant patients must get anti-rejection medicine to prevent the body from rejecting the foreign organ.

DNA

When we look closely at our planet, we find order is being created from what initially appears to be randomly distributed elements. The soil which contains all the nutrients necessary for life is basically inert. These elements are continually being converted to order (life) and decay (entropy). From the elements of the earth, man is born, grows and reproduces.

DNA is the blueprint of all living things. It contains all the information (intelligence) needed to mold the elements into a tree, a flower, an animal, an insect and a human being. Such intelligence cannot be of the soil but from some external source. This is because soil, by itself, is inert. The DNA cannot develop without a reference because a reference is necessary to develop order from randomness.

Everything we are and will be is programmed into our DNA. It could not have happened by chance or randomly, as suggested by evolution. The probability of that happening is zero. The entire theory of evolution implies randomness or chance selection which is at odds with the order seen in ourselves and in every living thing around us.

DNA is designed to conjugate only with that of its own kind. It rejects all others. This is contrary to what would be expected from random behavior. In other words, DNA is consistent and is highly selective in its reproductive characteristics. Since DNA is so selective, it does not lend itself to random change, let alone a systemic change from a monkey to a human being which would suggest cumulative positive changes.

Although similar, the designs are uniquely different between these two DNAs. Any similarities only indicate a common designer (creator).

The only mutations we see in DNA are degenerative in nature resulting in malfunction, thus indicating damage to its structure. This may manifest itself in physical deformity or breakdown of the body's immune system, both of which are negative in nature. These occur because the DNA (or genes) is now defective and not functioning as designed.

We typically take our body function for granted without considering how finely tuned the body's design. The human body is a complex integrated system. Its organs work together as a unit with each performing its programmed function. It needs all organs working in harmony, performing their specific functions, for the whole to be efficient.

The body knows when something is wrong and immediately tries to correct the defect to compensate for any adverse effects. This is automatic and indicates intelligence.

If we get an infection, the body immediately sends antibodies to the site to fight it and, in most cases, is successful. This is a continuous process in our bodies of which, most of the time, we are unaware.

It would have been miraculous if life had developed from a random process, but even more miraculous if something as complex as the human body was constructed randomly or by natural selection. Without intelligence, how would the body know that the design of its organs was now optimal and changes should stop? How did it know that a convex lens in our eyes produces an inverted image and so the brain needed to invert it for the image to be accurate? How would it have known that the tissue of the lens needed to be transparent and invisible to us? We know that anything other than transparent are defects, such as cataracts.

Even the process of growth is highly complex as the entire body has to grow in harmony. How many attempts did it take before it recognized that this harmony was necessary and how could it tell subsequent reproductions to make this change?

How did it evolve a brain to control all body functions correctly without knowing the entire design and processes of the body beforehand? It even developed a skull to protect the brain realizing that it was a very delicate structure that needed to be protected. Was this also an evolutionary process?

Why can't we find a skeleton without the skull if the body with a skull evolved and survived by natural selection?

Which came first, the chicken or the egg? Which is more confusing? Is it if we say the chicken evolved from the egg or was it the other way around? If we consider it to be some external intelligence that created it, there would be no confusion.

The egg contains all the information needed to make a chicken. It is an ordered system and so could not have developed randomly. The entire egg would have had to be fully designed before it could be closed in the shell, with no other changes needed. It would have had to have known that the information needed to make a chicken was now all there. There had to have been a reference and external intelligence. This could not have been random.

All of these questions point to an external intelligent source that designed and created living things. They could not design and build themselves because there is never enough information or knowledge at any stage of development to make these positive changes and fix them in future reproductions. The process of evolution would have needed to know the 'big picture', what was right and what was wrong and have the ability to select the sequence of positive changes. THIS TAKES INTELLIGENCE! INDEPENDENT INTELLIGENCE!

DNA not only designs the body but instructs it throughout its life on how to grow and reproduce. What degree of intelligence does that require? Certainly not some random process. This takes complete knowledge and understanding of the raw materials and how to use them. The raw materials being the dust or elements of the earth- the food we eat. It seems apparent that all of this was designed and created by an external intelligence. Not only did this source create it all but it was planned down to the very last detail.

Life is a continuous process. There is no time to stop and change design. All changes have to be made while the process is ongoing. It is much more difficult to try to make changes under dynamic conditions. It is much easier to stop and make the change and then restart. Life does not have this luxury and so it would seem more rational or logical that the basic design was complete before it was initiated. Any changes resulting from environmental interaction were also built into the original design (adaptation).

The Eye

When you look at one feature of the human body, the eye, do you recognize the degree of intelligence it requires to create such an organ?

We live in a three dimensional world as far as matter is concerned. (Time would be the fourth dimension but, for what we are considering, the first three are adequate). We have two eyes which are ideal for three dimensional viewing. If the number of eyes we have was randomly selected, we could have had one or three eyes or more. It would be viable for us to have any number of eyes. Remember there is no fixed reference, so there could be any number of eyes. But humans have only two eyes and have always had only two eyes. If this were not the case, we would have found human skulls with one or multiple eye sockets, not just two. This indicates that the creation of eyes was by design. This is the only logical conclusion.

Darwin developed the theory of evolution after observing differences or mutations in species, which resulted in their selective survival, in a changing environment. This is adaptation, which is built into the DNA and genes because the designer was aware of these possible environmental challenges giving the genes the ability to make these adjustments.

How can one explain life, let alone life responding to light, to form an eye. Then realizing we need two to be able to judge distance or increase viewing angle. There was no trial and error as we see no evidence that this was the case. All human skulls show that we have always had two eyes and two ears, located at the most ideal places on our bodies. This is also true for other mammals, birds and fish.

Evolution, by its current definition, is a random process. Only the end result is selective. From a random process one cannot expect to develop an ordered system. Even if nonrandom, this does not translate into being ordered because, in this case, nonrandom defines the ultimate result and not the process itself. Only the resulting elimination or survival of a species is nonrandom (this is more fully explained under the section relating to 'EVOLUTION').

The Blood

Apart from the brain, which is the control center of the body, the most important organ is the heart. The heart is responsible for pumping blood around the body which is how life is sustained. The blood carries oxygen and food to all cells in the body. This is essential to life. It also carries the antibodies that fight foreign matter that attack the body. It keeps all parts of the body alive and healthy, including the brain.

Man has always known the importance of blood to the body. This has given rise to superstitious beliefs that blood has some innate spiritual or supernatural power. Some tribes drank blood and sacrificed their people to appease the gods so that they may gain favor with them. They realized that blood is precious and essential to life.

As indicated above, blood flows continuously through the body as it carries precious food and oxygen to each cell in the body. The sap in plants perform the same function as blood in an animal.

The function of the blood and sap demonstrate that the process of life needs continuous renewal to grow and maintain itself. It is constantly fighting against entropy and cannot afford to put its guard down.

Blood is pumped by the heart but the heart needs blood to survive. The brain also needs blood to survive. The most important organs and all others need blood to survive. This means that blood is the source of life. It needs to be circulated continuously to sustain life. In a plant, sap needs to be circulated continuously to sustain the plant. Earth needs water circulating continuously to sustain life on earth. Do you see a theme developing here? Life needs continuous sustenance otherwise it dies. This is life as we know it on earth. It is a continuously maintained, ordered process. There is a fundamental interdependence of all living elements in the loop.

Death is an integral part of the life process. At a certain point our defense system begins to break down. This is a function of age. Death of natural causes will occur at some point in our lives. The average age is between 70 and 80 years. However, there are diseases that are capable of fatal attacks on the human body resulting in premature death.

> (Jesus died on the cross giving his precious blood so that we might live)
>
> Bible quote - (Mark 15:25)

Eventually life will fail, because death is inevitable, but the reproductive process produces new life and the cycle continues.

Design of the Human Body

When you look at the human body, do you think you could improve on its design? I would imagine that there are many people who would say they could. Let us look at examples of possible improvements.

The Hands

Most of what we have to do with our hands requires only one. Two hands are the optimal number for lifting heavy weights and for balance. But then, if there were three hands, where would you attach the third? If it were placed next to an existing hand it would be redundant. If placed in the middle of the chest it would be in the way.

When we look at the design of the body, as a whole, it is the optimal combination and location of all the body parts. Now, should we be led to believe that the design came about by evolution without the designer knowing and understanding the intended ultimate function.

The Heart

We may also say we should have two hearts, since the heart is so important. But one is enough if we take into consideration that it is protected by the rib cage and is designed to last the lifetime of the human body. If there were two hearts the designer would have had to take into consideration blood flow issues which becomes more complicated with two hearts. The fact is, the body's design is adequate, efficient and optimal for its purpose.

Medicine

In medicine, the reference is the healthy human being. One of the first things a doctor does with patients is to take their temperature. For a healthy person, the temperature should be about 98.6 degrees Fahrenheit. If it is too high, the person has a fever. That means that something is wrong. If too low, that is also a problem and has to be corrected. We should keep in mind that 98.6 degrees Fahrenheit is the average acceptable normal temperature but small variations above or below this temperature are considered within tolerance.

What we have learned is that the human body operates within certain parameters. If it goes outside these parameters we know something is wrong, we are ill.

Other parameters include blood pressure, blood composition, urine composition, skin color and texture, body weight etc. As we gain more knowledge about the human body, we develop a more comprehensive list of parameters that are the norm. Any abnormal variation needs to be addressed.

Since the human body is an ordered system, there are certain norms that we look for in a healthy body. These remain fixed and so are the fixed references.

Having reviewed the order we see in nature and the universe, we will now look at evolution to see if its theories are compatible with the principles of order.

Evolution

I will define the evolution of man using the two interpretations or theories with which most of us are familiar. Evolution, in general, also follows the same principles.

1. Darwin - The changes, over time, in species as a result of genetic differences giving one an advantage over the other in a changing environment, with the eventual disappearance of the weaker or disadvantaged species. This change may be random or nonrandom.

2. Primordial - The random development of the life process, from what we call 'primordial matter', to life as we know it in its current manifestation.

Note: Neither of the above theories suggests how life began. Life can only develop from something already living.

Interpretation 1. - Darwin's theory of evolution.

Darwin seems to suggest that DNA gets more complex with time, with no explanation as to the mechanism of this change. In the process of NATURAL SELECTION, as Darwin postulates, the inferior or disadvantaged species eventually die out as a result of adverse environmental conditions which they are unable to overcome, whereas the more resilient species are able to survive and multiply. This would also imply that the change is cumulatively positive if we look at fossils of lifeforms connecting our human species to its current form. Species are defined by DNA and so, when species die out, they become extinct.

Interpretation 2. - The primordial theory.

The primordial theory is even less plausible since it not only does not explain the mechanism of the change but also suggests that there were billions of cumulative positive changes that seem to randomly occur.

Let us first look at the nonrandom process of evolution.

Nonrandom Evolution

We know, by observation, that life is not a random process. It is the result of an ordered process controlled by the DNA of each living system. We also know that healthy DNA resists change.

Nonrandom evolution may be said to be due to changing environmental conditions resulting in the elimination of some species and the survival of others based on DNA or genetic differences, giving one an advantage over the other. This would be considered a lateral change as this is only about survival and not necessarily considered an improvement. It is, however, selective, hence nonrandom. This is the only evidence of nonrandom change, as the mechanism of the internal change would be considered random, based on the Darwinian or the primordial theory, since the basic change mechanism is not defined and there is no explanation as to how or why it occurred.

The diversity we see among the species indicates change based on genetic differences. If genetic changes occur first and then the environmental changes follow, this may result in mutation or elimination of the weaker species.

But what would initiate a single change in the DNA or genes of a living organism, much less several positive changes, to develop a significantly superior species?

For this to occur, there would have to be an intelligent source setting or recognizing the fixed reference (DNA) and influencing these changes, in a sequential manner, in harmony with the reference (DNA). This cannot be a random change mechanism because it results in cumulative positive changes. Time also does not explain this change as the longer the period over which this takes place, the higher would be the probability of a truly random outcome. Over time, a random process does not improve on itself and, the more time elapsed, the more the process would tend towards truly random.

The believed evolutionary change from a gorilla to a human being could not have occurred without an intelligent source influencing the change, using the fixed reference (DNA) of that species as a guide in order to ensure a cumulative, positive outcome. The only connection between the gorilla and the human being is that they both have the same designer. They are actually two separate designs by the same creator.

Throughout this entire process, intelligence, whether built in or with direct intervention by an external intelligent source, must have been involved.

If the initial reference is not actively included, there is no guarantee that changes will be compatible with the system or go in a positive direction. Intelligence is built into DNA, but how did it get there?

How would the process have known when the goal had been met and no other systemic changes were necessary? Human DNA appears to have known that what it built was good and no further changes in the structure were necessary. Healthy DNA only allows changes compatible with its design and resists other changes. This shows a form of intelligence.

If this intelligence did not originate in the structure, it must have come from an external source. As far as we know, matter cannot develop intelligence by itself and so we can conclude that this intelligence is from some external source.

The original life process could not have been random because it developed an ordered system. Randomness can only develop order if instructed to do so by an intelligent source. Evolution would have to be an ordered process to develop or complement an ordered system.

Now, an ordered process must be of an intelligent source. Intelligence defines order. Evolution, as it manifests itself, must therefore be of an intelligent source. It cannot be defined any other way as it enhances an already ordered system.

However, neither of the above theories includes intelligence as the guide to the development of the evolutionary process. Nonrandom evolution, as it is currently defined, has no intelligence, it occurs by 'natural selection' of the species with the advantage in a given environment. It cannot reason. It is, therefore, not feasible for such a process to successfully initiate or continue to enhance an already ordered system.

Evolution, as currently defined, cannot explain how life was initiated. It also cannot explain the cumulative positive changes in the gorilla's DNA to become human unless it was guided by intelligence.

On closer examination, we would conclude that the amount of error associated with the design and creation of the human body is infinitesimally small or nonexistent. There are checks and balances inherent in its design to ensure minimal error. Where error is found, it is a result of an adverse interaction with the external environment and possibly from the system itself as a result of 'GOD'S CURSE' that has affected ordered matter, in various ways (see Bible quote below). One should also keep in mind that there is one reference in the Bible were JESUS said that a man was born blind, not because of SIN (or the CURSE) but because GOD wanted to demonstrate his power over sickness by JESUS performing a miracle to make the man regain his sight (Bible quote next page). We were originally, perfectly designed, but we now live in an imperfect world. Our design reflects an intelligent creator.

If we look at genetic defects such as Down syndrome, this occurs when an abnormal cell division causes extra genetic material from chromosome 21 to be formed. This is not how it was originally designed, but there has been some form of malfunction which results in extra genetic material being generated.

Based on the above reasoning, defects may either be from adverse interaction with the environment or a defect of the system as a result of the general effect of SIN on matter in an ordered system. Matter still conforms to the laws of nature, but SIN has not only affected our mind but also matter based on its effect on an ordered system.

Bible quote (Genesis 3 - GOD curses creation)
Bible quote (John 9:5- Jesus heals the blind man)

Random Evolution

Here, it is also implied that changing environmental conditions result in the elimination of certain species based on DNA or genetic differences making one more vulnerable than the others. If we consider this, it is also selective, as the weaker or disadvantaged species is eliminated. Only the internal change process could be considered random and so the same argument applies as was discussed in non-random evolution.

But again, what would initiate these internal changes? If these changes are randomly initiated there is an equal probability of each resulting in a positive or negative outcome. In the long term, things would not improve or result in a cumulatively positive outcome. Species would therefore not improve from their initial form. Also, as the system gets more complex, the probability of getting a positive outcome is continually being reduced as there is now a decreasing chance of getting a positive outcome. The reason is that the process would now be getting more selective.

The only way to ensure cumulative positive outcomes is if there is an intelligent source in coordination with the fixed reference (DNA), of that species, influencing the changes. Otherwise, changes may be detached and not compatible with the rest of the system. Changes must be overseen by an intelligent source to ensure that they are complementary. This could not be a random process.

Entropy works against life. Yet, life is designed to combat most of the day to day encounters that threaten it. These defenses are built into the system. This could not be a random process because life has a systematic way of dealing with environmental or internal attacks. It appears to develop a logical and rational line of defense, not a random one.

Based on what we see in the design of the human body, the designer would have to have had a plan. Only a source with intelligence can create a plan. After developing the plan, it had to have been executed in a step by step manner until completion. Each step would have to have been precisely followed.

When we, as humans, design and make things, we develop a plan and must use some form of reference so that we know where to start and the direction in which we should go. We have learned, from experience, that this is the only way this can be successfully done.

We are using the same building blocks, in the things we design and make, as those used in nature, as these are all we have available to us. We, therefore, must obey the same laws that govern matter. Nature and man must then follow the same fixed guidelines to achieve the same results (an ordered system).

To begin, there has to be a fixed reference, with all other references defined by the initial reference. This is the only way to coordinate the points or components in an ordered system with multiple subsystems. It takes knowledge of the complete system as well as sequential execution of the plan to accomplish this task. This could only be done by an intelligent designer.

A random process is the opposite of an ordered process, with a diminishing chance of a continually positive outcome. A purely random process cannot improve on itself. If it is biased towards the positive then it cannot be defined as being random.

Every ordered system has a fixed reference and follows a specific sequence of development. In addition to sequence, the designer must have some concept of the end product. Otherwise, how could it know when the system was complete and operating as designed? In other words, it must have intelligence. This principle is universal in its application. This is evident in creation and in the way we function, being made in the image of GOD.

> Bible quote man made in GOD'S image
> (Genesis 1:26-28, Solomon 2:23)

In every material ordered system, there is a degree of tolerance, with extremes on either side of the normal distribution (Bell curve - tolerance is the allowable amount of error). This measuring standard was set by us as we have found that many natural occurrences follow this pattern. The extremes of the distribution, although not normal, are considered natural. An example of this is autism, where, although brain function is not normal in a person with autism, with proper treatment, the brain can be made to perform within normal parameters. This indicates that we are designed to function within certain parameters (ordered) and outside of this we do not operate in harmony with each other.

There is no evidence that the process of evolution, as we understand it, recognizes, much less uses a fixed reference and so it would be unable to create an ordered system.

Evolution can either be influenced by environmental (external) changes or internal changes. Evolution is 'LOST' without a fixed reference. It appears to have random references or no reference at all. Without a fixed reference, how is it able to know which direction to go? How can it, therefore, design a complex structure like the human body. Such a system requires the coordinated sequence of complex, positive changes which can only be accomplished by an external intelligent source using a fixed reference.

In the case of the human body, it would take infinite intelligence to design, create and incorporate self-maintenance, as well as the ability to reproduce itself.

If we look at our daily lives, our ultimate intent is to create order. Disorder and randomness are nonproductive. In order to accomplish a goal, we first need to have a plan as to what we want to accomplish. The next step is to determine the starting point, then take sequential positive steps until completion. This is not a random process and can only be accomplished by an intelligent mind.

How can someone examine the human body and conclude that it was put together by a process that was unaware of the end product but ended up with something that, even with our intelligence, is far beyond anything we could have conceived much less created. There is also the fact that it can reproduce itself.

Any ordered system that we encounter or create is planned and then executed. This is not in line with the principles of evolution as we know it. In fact, evolution is the only process that we know which is said to have developed a complex ordered system with no plan and no concept of its intent.

How could we expect that evolution could start from a primitive living organism (primordial) and eventually change into a human being, by a random or nonrandom process without intelligence. All odds are against this type of change as entropy is one of the main processes that would prevent this from happening.

This would mean understanding the laws of nature that govern matter. Evolution is not rational or logical. Some outside intelligent source must have intervened.

To create a human body, it would take billions of sequential, cumulative, positive developments which could not result from a random process. It would take a positively biased process with intelligence to accomplish such a feat. Evolution, as currently defined, exhibits none of these characteristics.

To believe that life evolved by the process of natural selection does not explain what we see in nature today. What we see are the completed products of an ordered developmental process (ordered systems). Natural selection is only the process of elimination. What we see are ordered systems both in life and in the manifestation of the universe. No random or nonrandom selection process could have developed either of these systems. It took infinite imagination with a detailed plan, executed in flawless sequence, monitored at every phase, and still is, to the most intimate degree.

This takes infinite intelligence and infinite knowledge that could only be accomplished by a GOD.

Creation was planned for man.

Bible quote - (Jeremiah 29:11, Ephesians 2:10, Philippians 1:6)

A Bird's Flight

Evolution implies that birds evolved wings that gave them the ability to fly. This conclusion does not make sense because, for this to happen, there had to have been an intentional effort with successful positive outcomes to develop a viable aerodynamic design.

Firstly, why was it necessary for any living thing to fly? There is no evidence that this was necessary and so it must have developed randomly as the process of evolution would probably suggest.

Now, if this were a random process, over a period of time, for every positive development there would be a negative, and the longer the period, the closer the process would approximate to truly random- say millions of years. At any point in time during its development, there would only be a few positive feasible events and billions of negative events that could halt the process. The process would have to have been guided by a plan as to which direction to go and what sequential steps to take to attain flight.

Some of the things that would have to have been successfully worked out would have been the following:

> The need for wings
> The wing design
> Location of wings
> Balance of the two opposing wings
> The design of the bones in the wings
> Where the joints needed to be
> What motion would be needed for flight
> The location and strength of the muscles
> The need for feathers
> The design of feathers

The material needed for the feathers

The size and shape of the body relative to the wings

(Wings are so unique we copied the aerodynamic features of the design for our own airplanes.) We could not improve on them.

These are just a few of the obvious things that would have to be considered for a successful flight. If one of these was wrong or in the wrong sequence, flight would not be possible. Even after evolution got everything right, the bird would still then have to learn to fly and communicate this information to its offspring.

If we look at birds flying thousands of miles and finding their way to the same destination year after year, how do they manage such navigation? Even if we do not find out exactly how they do it, we know that they must have some form of a fixed reference. There has to be something that remains constant that they can use as a reliable guide.

What I am saying about evolution is that it got all these things right by trial and error. This is not possible unless the end product is known and there is an intelligent mind to guide the process. Evolution is blind without a reference. It could not have accomplished such a feat.

Below, I have discussed a possible explanation for the changes we see in fossils of humanlike bones and those of other species, as the species evolved.

Alternate explanation of Evolving humanlike and other Fossils GOD could have set creation in motion so that all the evolutionary changes would occur in sequence, over time. Like the development from a child to an adult and then old age. Sudden and dramatic like those changes at puberty and a chrysalis turning into a butterfly.

From 'homo erectus' to 'homo sapiens' as well as all the previous evolutionary changes associated with human beings. These could have all been programmed. One certainty is that these changes had to have been guided by an intelligent designer.

Initially, all that He needed to do was to create the basic DNA structures for each living thing and program changes to occur after predetermined periods of time, to adapt to changes in the environment. The basic structures would include man, animals, fish, birds, reptiles, plant life and micro-organisms. Programmed changes would create new species, with some becoming extinct over time and others multiplying and continuing to survive to what we see today.

There was one creation but it is probably much more complex than how it initially appears. GOD could have programmed man's DNA to change at specific points in time, to increase the capacity for the acceptance of more sophisticated knowledge processing and to change in appearance to what we see today.

This would mean that creation is much more complex and dynamic than what first meets the eye. This would present a much more complex picture of GOD and His power to create and maintain an ordered universe, but all within his capabilities. He is 'ALL POWERFUL' and 'ALL KNOWING'.

But, this should be apparent when we examine creation. The changes in a human body from birth to maturity and then old age. These changes are all programmed to occur at set times. Eventually, life ends in death.

GOD could have planned that at certain thresholds, systemic changes would occur in the structure of the DNA, so that a new and improved form of life was established. This would continue for a specific season, after which another change would occur to further improve that life form. Still, there may be some that are not programmed to significantly change such as the cockroach.

The average human life time is about 70 years. But, if we look at the creation of the universe, its life time may have been programmed to be billions or trillions of years. What we see could be the unfolding exactly as GOD planned. All of which appear to be evolutionary changes could be programmed changes like the chrysalis to the butterfly, the egg to the chicken, and the human body transformation at puberty. The cycle, however, is much longer.

GOD is capable of all this and much more. We cannot even begin to imagine His power and majesty. Anything you can imagine cannot even begin to describe our GOD. Our imagination does not have the scope to comprehend the power and intelligence of GOD.

We see these patterns everyday in our lives and in everything around us. This is all in line with GOD'S power and capabilities.

Adaptation

Adaptation is built into an already ordered system to adjust for environmental changes that threaten the species. It was designed by an intelligent source that anticipated these changes because the source could see the 'big picture'. This is another manifestation of planning.

Adaptation is more in line with the small changes we see in living organisms to compensate for environmental changes that would otherwise threaten their existence. This has been mistakenly extrapolated to represent the systemic changes that are necessary to develop a living ordered system.

Adaptation may be feasible for small changes but not feasible for systemic, positive changes while still maintaining an ordered system. Even so, it is of an external intelligent designer and is built into the system design.

Adaptation did not create order. Evolution did not create order. Intelligence defines and so must have created order.

Daily Human Activities

If we look at our daily lives, our ultimate intent is always to create order. In order to accomplish a goal, we first need to have a plan of what we want to accomplish. The first step is to determine the starting point, then take sequential positive steps until completion.

Your daily activities start with getting out of bed. First, we have to recognize our position-that we are in bed, in a horizontal position. We have to be aware of the location of our feet and where they are positioned. This seems very basic but what we are doing is using references, being always aware of our current position and what next to do. This is all automatic. We know what muscles to move and in what sequence to move them in order to get to the upright and then standing position. What we do next will depend on our subsequent plan or intent.

We are continually aware of where we were, where we are and where we want to be. Only an intelligent mind with memory can recognize and understand or interpret this sequence. If we are unable to do this, we are lost.

If we look at an object, we are only able to recognize it because we have a prerecorded reference. Our brain works in an ordered manner with a series of references to which it has multiple or what some might call, random access.

When we have a thought, it may have been initiated by something we saw, heard, felt, tasted or smelled. It may also have been a thought with an origin we are unable to specify. In any case, the thought must have a reference. Once that reference has been recognized the brain makes the association with the connecting information already recorded. Otherwise we are unable to interpret the thought or act on it.

Reproduction

The human body can only reproduce itself. It was designed using the DNA blueprint which was specifically designed for humans. Similarly, DNA for other living things can only reproduce themselves. This it true for all species.

In the same way, when the architect designs a building and draws the plans, it is only for that or buildings of the same design. Another plan has to be drawn for a building with a different design. If it is subsequently modified, this modification has to be done by an intelligent mind using the initial blueprint (reference) as a guide.

An analogy to this would be annual improvements made in the design of an automobile. Each year the model changes and improvements are made. The improvements do not occur by themselves. Design improvements are made by the design engineer, the intelligent designer. Automobiles evolve but only at the instruction of the intelligent designer.

The reproductive process follows the instructions of the DNA of the species. It does not randomly change.

Parent And Child

There are unique similarities between parent and child. These similarities include both physical and mental traits. In some cases, the visible resemblance is so striking that it almost guarantees a close relationship between the two subjects. This similarity is also some-times evident in the extended family.

This is an example of how nature works. It is how things mani-fest themselves indicating close relations and connections, based on these similarities.

The only evidence we have to work with is nature, the universe and some understanding of how it all works. Now, wouldn't it be rational to conclude that, what we observe in nature shows evidence as to its origin. This means that there are traits in nature and the universe that manifest their origin. I think the most fundamental of these is basic 'ORDER'. Before these traits can be reproduced, there has to be an ordered transfer of information.

In electrical waveform amplification (an audio amplifier) the goal is to amplify the signal without introducing distortion. The amplifying circuit is carefully designed to accomplish this goal. But, even though this is the goal, there is always some evidence, in the amplified output, of the characteristics of the amplifying circuit. In this case, it manifests itself as distortion or some unique character-istics of the parent circuit that are evident in the output, even to a very small extent.

The laws of nature are also similarly designed to reproduce and maintain a specific theme or themes, but leaving some indication or evidence as to their origin. The product is intentionally designed for a specific purpose but contains evidence of its source or origin.

An example of this is a product made on a lathe. If we look at the macro structure of the surface, we see the grooves made by the cutting tool in the form of a close spiral on the finished part. This indicates the process used for making the product was 'turning', on a lathe.

ORDERED SYSTEMS FOUND ON EARTH

W E WILL NOW examine a number of ordered systems that man has developed as well as ordered systems of other intelligent sources. To reinforce this theory, each of these systems must have a fixed reference.

Here are some examples of ordered systems of intelligent sources other than human:

> Birds flying in formation
> A school of fish swimming in unison
> Ants on a mission
> A swarm of bees in search of a new site to build a new hive
> Bees in the process of building a hive (honeycomb)
> Animal pacts living and hunting together

They all work together as a team, as one. This is order. It demonstrates intelligence. When we look at our world, we will see how references apply in every aspect of our lives.

Firstly, we must remember that to be ordered, events must occur in a set sequence. To set the reference and monitor the sequence, there has to be intelligence.

A Bird's Nest

Let us start with a bird's nest. The end product is a system that will offer protection and a home for its offspring. The bird starts by gathering twigs and leaves, beginning at the base and interweaving them to form the sides until it is complete. The twigs are interwoven to form a circular interior, starting from a single point and connecting the whole together.

All birds have a similar plan which they execute in sequence. The nest must be of adequate size to hold the eggs and the mother bird while she sits on them until they hatch. She must take into account the number of eggs that she may lay and the maximum size of the birds that will occupy the nest, before they leave. The nest is necessary to keep the eggs together in a safe place for them to hatch and the young birds grow, until they are ready to take flight and begin life on their own.

The nests of each species are consistently similar. In some cases it is possible to look at a nest and identify the species of bird that made it. Each nest is made from an instinctual plan and executed sequentially from start to finish.

Such a system requires a fixed reference because there are certain critical features such as the shape and the size that need to be consistent. The bird must decide where the base will be and where the sides will begin to determine the internal dimensions of the nest. It must therefore have a fixed reference from which to construct these dimensions relative to each other. In order to construct such a system, the bird must have the degree of intelligence, first to select a suitable spot for the nest, find the twigs and leaves, which will be the raw materials to build the nest, and then begin the process of building. It must also know when it is complete and ready to provide a suitable home for all the eggs.

The interior of the nest is basically circular. If we look at the geometry of a circle, we see that in order to make a circle, we have to have a fixed reference which is the center of the circle. The distance from the center to the perimeter is the radius and this is constant for any given circle. The bird has a mental image of the center and the radius or diameter and constructs the circle with these kept constant or relatively constant, as there are allowable tolerances for such a structure. In other words, some error is allowed.

It takes intelligence to construct a nest. It also takes considerable skill to weave such a structure. If you try to make such a structure you would find it very challenging and you may never get a finished product of the quality and structural integrity of a bird's nest. Yet, we have hands with opposable thumbs and a much larger and complex brain.

If you saw a bird's nest by itself and you did not know its connection to a bird, would you think that it was constructed by some form of intelligence? If we consider it an ordered system, then by definition, it would have to be of an intelligent source. By observation, we know it is an ordered system and by looking at the design, we see that it has a fixed reference.

A Bee Hive (Honeycomb)

Let us look at another ordered system- a honey comb. If one saw a honeycomb in a tree and did not see the bees that made it, we would believe that it was of an intelligent or ordered source. It takes intelligence, even at this level to put such a structure together.

Now, look at it from the bees' standpoint. They are given the task of making a symmetrical structure from a single material, which they produce, in order to provide a suitable habitat for their offspring and to store honey. In order to make a honeycomb, they have to start somewhere. This is the reference. Once they set this reference they must stick to it in order that each cell is in geometric symmetry with the adjacent cell.

There are many bees participating in this project and so the reference information has to be accurately communicated to all of them and they cannot afford to make a mistake. They have to work as a team. If they changed this reference while in the middle of building, the hive would be deformed and would not come together properly. It takes just one single change in the reference for this to occur and so, all subsequent cells, after the first, must be made with reference to the first. This is only a simple example but the principle is critical in the development of any ordered system.

Birds Flying In Formation

For bird's flying in formation, the lead bird is the reference. For bird's traveling long distances, the lead is rotated as this position requires significantly more energy exertion, whereas the other birds fly in the 'slip stream' made by the lead bird. Flying in the slip stream requires less exertion, making longer flights possible. The lead is rotated so that the lead bird gets to rest. When we observe this formation, for geese, this is a perfect 'V' indicating it is ordered, thus exhibiting intelligence on their part.

These are a few examples of intelligent sources, other than human, creating ordered systems.

Now we will look at the human being and how we have developed ordered systems for communicating, teaching, manufacturing and navigating, as well as some other familiar ordered systems we have constructed.

Any ordered system we have developed, in every case, it was done to simulate, describe or communicate aspects of the ordered system of which we are a part.

Remember, if we now look objectively at the big picture, in our effort to try to determine creation's origin, we must look at it like a detective examining a crime scene. We look for 'fingerprints' or patterns in creation that point to the character of the creator.

I will now review several disciplines, with which most of us are familiar, to highlight each of their fundamental components. The theme that will become apparent is that each, fundamental component has a fixed reference and complies with physical laws that do not change. This is an indication that there is some reference, of an intelligent source, keeping these laws constant or ordered. Things only remain ordered if they have a fixed reference. The reference must be set by an intelligent source that understands sequencing. This is the only way to orderly develop and grow from the initial reference. But first, I will start with man made ordered systems.

MAN MADE ORDERED SYSTEMS

Communication/ Language

Since we come into this world by way of our parents, we are directly linked to them. They display an innate desire to love and protect us. They take care of us when we are unable to take care of ourselves.

They provide us with the necessities of life and are the first to teach us the basics for survival. This trait is seen throughout the animal world of which we are a part.

After we gain awareness of being alive, one of our first instincts is to communicate with others. As a baby, we instinctually use the universal language which is crying, meaning, 'I am unhappy', a smile which means 'In am happy', and so on. We then begin to learn the common language. This is the first ordered system that we learn for detailed and accurate communication.

The universal body language is the way we first communicate without the spoken or written word. We all, as humans, understand a smile, a frown or other facial expression. Similarly, gestures like beckoning with one's arm means 'come this way'. Body language has meaning and we immediately recognize an expression of friendliness or rejection without the spoken or written word. Communication requires intelligence on the part of both parties involved. The universal reference for communication is body language but, as we get more sophisticated, we develop the spoken and written language.

The spoken language incorporates specific sounds in a set sequence and with specific meaning that is always consistent. Then, we go to school to learn how to read and write the language. Later, we learn mathematics, the sciences and the arts. This is the accepted practice. This is how we are prepared to take our place in society in the pursuit of knowledge and to perform a responsible role. Through all of this, we are trying to understand ourselves and our world.

If we examine language, we see that it is ordered. We develop specific sounds or combinations of sounds to communicate specific thoughts. Everyone in a community must use the same combination of sounds in the proper sequence to express the same thought, request or emotion.

With time, we learn how to communicate fluently. The better we communicate the better we will be able to accurately exchange thoughts and ideas. This is important as we need to operate as a community. Each of us is a part of that community and to make our best contribution, we must be in harmony.

Then we progress to the written word, since it is sometimes necessary to communicate with someone out of ear shot. We develop common references such as an alphabet, words and sentences. The alphabet represents specific sounds. A word is the combination of alphabet symbols representing these sounds, in sequence. The sentence is a combination of words expressing a thought, a wish, a question or a command.

In any specific language, these 'words' must remain the same for each specific communication or meaning that they represent. In English, a 'girl' is a young female human being. A 'dog' is a specific type of animal, as is a 'horse' or a 'cow'. We have to keep these constant otherwise we would be totally confused and language would lose its purpose.

Each nation has its own native language or dialect. Within that language, the words must have the same meaning. We take this for granted but what we have done is created a communication standard or reference for all in that group. In addition, the reference has to be fixed to avoid confusion.

As I have mentioned earlier, from the time we are born we begin to be programmed by our environment. We are gathering information and knowledge about our world. The only way we are able to do this is because we are intelligent beings. We are born with intelligence. We have a memory so we are able to recognize sequence and learn by a process of information or knowledge accumulation.

Information cannot be stored randomly, otherwise we would not be able to use it efficiently or possibly even access it. There must be a reference to which all individual bits of information are connected and logically stored and accessed. In other words, it has to be ordered. This is the only way we would be able to access, store and retrieve this information accurately and efficiently. We have also learned that information storage in the brain is redundant which makes it easier to access.

We can now see how important it is for consistency, in order to maintain accurate communication when using language.

Societies are continually inventing new words to describe specific new inventions, concepts and thought processes as we grow in knowledge and change. However, we still have to be consistent in their application to avoid confusion. Sometimes words disappear from our vocabulary if they are no longer relevant or necessary in our changing environment or society.

In all communication, consistency is critical as in a common meaning of a particular word or expression so that we accurately convey our thoughts. Accurate verbal communication is extremely difficult as sometimes we never know how our words are interpreted. Cultural differences (accents) can result in different interpretation of words, even if this is not the intent. For the most part, this is the best we have and so we should constantly hone our language skills.

In our world, there are several languages which make it difficult for us to communicate verbally with everyone. Wouldn't it be nice if we all had a universal verbal and written language. Then we would have a universal communication reference. Different cultures could still have their ethnic languages or dialects that are unique to them.

Our 'heart' is that which accurately represents our thoughts. Language, at best, is only a close approximation of the thoughts we are trying to convey to others. Sometimes we say things that are the opposite of what we are actually thinking. This occurs when we practice deceit. GOD knows our 'hearts' and so He cannot be deceived.

Mathematics

We take learning mathematics for granted. It is expected of us. But why is this necessary? What we are doing is simulating natural events or occurrences. We need a standard or reference to communicate what we see in nature- two oranges as opposed to one. How about 100 oranges or more! How do we communicate this to someone else? We use language and symbols (numbers) to represent mathematical principles so we can communicate linear or even continually changing conditions as in the case of calculus. We learn to count. But first we have to have a reference and that reference must be fixed. We call this reference 'ZERO'.

If some people changed this 'zero' reference to 'one' and everyone else kept it at 'zero', all their calculations would be incorrect with reference to all the others in the group. So we can see how important it is for us to keep the reference fixed for everyone. We cannot be in harmony unless we have the same reference. Harmony means ordered as opposed to random or disordered. We are only able to simulate natural occurrences with this mathematical system because nature is also ordered. Nature came first with order and mathematics was designed to simulate it. Only order can simulate order.

As we progress from simple linear calculations to hyperbolic functions, the principle still holds. Hyperbolic functions, such as sine and cosine, simulate wave forms such as light and sound waves. Exponential functions simulate population growth and other real life occurrences that start slowly but dramatically increase over time or vice versa. (see diagrams below)

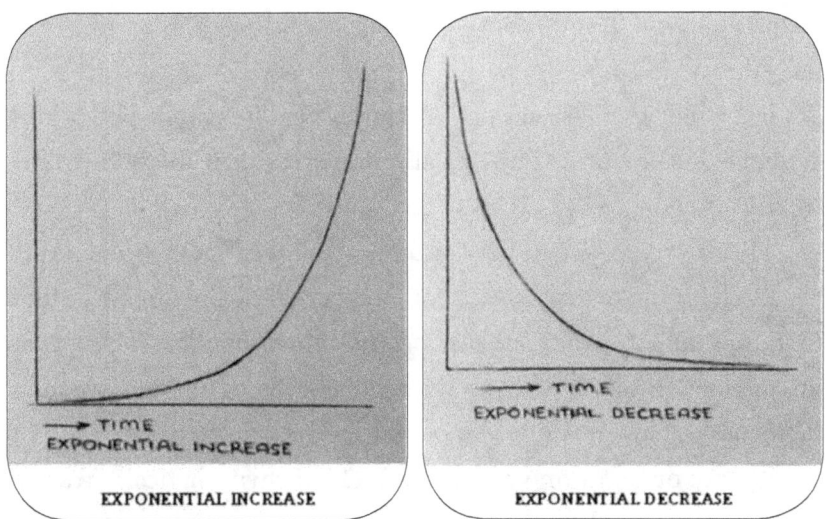

Population characteristics are represented by the (Bell curve on page 81). This is discussed later including standards such as mean, and standard deviation. In every case, we use these equations to simulate nature. Also, in every case, we need a fixed reference. Intelligence is a given in setting these references.

Foot Note

The following is a simple example of how we calculate unknowns from an equation with other known factors. This is a typical application of a linear mathematical equation.

Look at the simple equation below:

$$C = A \times B$$

To determine the value of C we must know A and B. To determine how C changes with respect to A or B we need to keep either A or B constant and vary the other. If we make B= 2, then C= 2A.

In this equation C and A bear a linear relationship. C is always twice the value of A.

If we plot these values on a graph, we get a straight line with a gradient or slope of 2. The line also passes through the origin since when A is equal to 0, C is also equal to 0.

This is very basic but the concept is true in all such calculations.

In defining nature's behavior, we develop equations like this that approximate what actually occurs. There may be some errors or small variations but these are negligible based on the generally predictable characteristics obtained from the results.

We say one and one equals two. This is hypothetically true. If we were to apply this to an actual situation, we would say one egg plus one egg equals two eggs. However, no two eggs are the same. Technically, one egg may be larger than the other which would make them not equal. Or there may be color or shape variations. But for all practical purposes, one egg plus one egg equals two eggs.

When we design something, we take into account that there will be variables that we will not be able to fully control. So we develop 'tolerances' to see how much error can be tolerated without affecting the intended function of the product. Tolerance is discussed later on in this book.

Manufacturing

I am a Manufacturing Engineer by training. I obtained my first degree from the City of Birmingham Polytechnic and my Masters Degree at the University of Birmingham, England. My major was Manufacturing Management and my job in industry was to coordinate the various processes and disciplines in a manufacturing plant to ensure that the finished product was made on time and to the design specifications. These disciplines included Mechanical Engineering, Electrical Engineering, Civil Engineering and Electronics. This gave me the opportunity to work in and also visit a variety of manufacturing facilities. In so doing, I became familiar with many manufacturing processes, from General Manufacturing to Medical, Technology and Aerospace. In addition, I was employed for 35 years with an insurance company as a Risk Control Consultant. This gave me access to many manufacturing facilities looking at quality control procedures from a product design and product safety standpoint.

All manufacturing operations have one main thing in common, detailed manufacturing procedures including quality control at all stages of manufacture. This is essential for producing high quality products in a predictable, repeatable manner.

Have you ever made something, anything? Do you understand the precision it takes to produce mechanical parts or parts for furniture so that the finished product is functional and aesthetically pleasing?

I have always enjoyed several hobbies including model making, oil painting, electronics and woodworking. With this combined experience, I know what it takes to design and make a product. What it takes to select and shape raw materials into a finished product; an ordered system.

I am now retired and have been for more than six years. I therefore have the time to review the knowledge and experience that I gained throughout my life and put it all in perspective.

I have learned that the materials must be right for the application in order to make a quality product. You have to understand the properties of the various materials used in a project so that they are compatible in the finished product. It also takes the relevant knowledge of production processes and the proper sequence throughout the component manufacturing and assembly processes. If the final step is incorrect, the entire project is at risk even though you have done everything else right.

It takes intelligence to keep track of the sequencing so that it is carried out in the proper order. In some cases, even the timing is important- not too fast and not too slow. The process is far from random. In fact it cannot be random as, by definition, it is an ordered process.

Frustration occurs when you are working on a project and it is not progressing in the way you anticipated. This happens because your production procedures are incorrect or you are not following nature's laws.

If you are fabricating something from steel, you must use cutting tools that are harder than the material being cut and also not brittle that it breaks under the cutting force. Cutting speed is also important, otherwise, the tool may overheat and lose its hardness and become ineffective. Cutting speed may also affect the tolerance and, if this is critical, we need to control these variables as much as possible to obtain consistent results.

What I am trying to say is that we have to be consistent in order to have predictable and repeated success. This process is not random but follows a set sequence that must be closely followed to produce the desired finished component. The same procedure is required for all components. The final assembly of the product is an added dimension where there is an additional possibility for error, even if all components are within the required specifications. This is because things like alignment and compatibility must now also be taken into consideration.

To produce several accurate components that make up the finished product, a system for quality control must be in place. This is critical to ensure that quality products are consistently made.

Instead of machining, an alternate method would be to melt the steel and cast it into the desired shape. This is a different ordered process to achieve a similar end product. The desired end product is known and the manufacturing sequence is formulated to get it done. This takes intelligence. Throughout the process, accuracy is continually being monitored in order to produce a quality product. We are also aware when the end is reached and where the process must stop.

Similar rules apply when working with various other materials including wood, fabric, plastics, chemicals etc, which all have different properties.

There are always set guidelines with which one needs to be familiar in order to succeed. If any one of these steps is missed or selected out of sequence, you will be unsuccessful.

Looking at the big picture, anything that is made or produced is done using set guidelines from start to finish. In all cases, only the raw materials available to us are being used. These are the same materials, with the same properties, used to make the universe.

Some people may be gifted in design and fabrication and are therefore able to make an original design and finished product without much effort. This is not the norm. It takes talent and experience to accomplish such a task. One should never underestimate the challenges such a task presents.

CNC Machining (Computer Numerical Control)

One uses the same principle of planning and setting a reference in CNC machining, to accurately make machined parts for any application. The machine tool control program must be given a fixed reference point typically(0, 0, 0), before it can accurately machine a part. These are the coordinates of the fixed reference point. The cutting tool must always know its location with reference to this point. This is the only way it can recognize its current location or coordinates and know where next to proceed.

This is also the technology used in 3D Printing. Instead of using a cutting tool as in CNC machining, the program now controls a type of spray nozzle that ejects the product material onto a flat surface (grid). Based on the dimensions of what is being produced, the nozzle ejects the material to the corresponding dimensions starting at the base and building up in the vertical direction, to exactly duplicate the three dimensional coordinates of the product being made. What it does is build the product slice by slice, until complete. This is the process of integration as is learned in mathematics. The smaller each incremental step, the more accurate will be the finished product.

Before CNC machines were invented, standard references were used to measure dimensional accuracy as the machining process approached the dimensions of the finished product. The machinist needed to have these standards at his or her machine to measure the accuracy of each part as it was machined. These standards are made from a very stable material to very high dimensional accuracy. They are, therefore, suitable for use as references when measuring to high degrees of accuracy. These standard references include, micrometers, vernier calipers, and similar high accuracy measuring instruments.

The first straight edge could have been a container of water with the surface as a reference. The surface of the water would be perfectly flat under gravity and so would be a good reference. Another reference could have been a string suspended with a weight on the end or a string pulled tautly between two fixed points. We had to start somewhere, but once we established the basic references, we were able to improve on them to obtain more accurate results.

Tolerance

Every material ordered system must have some limits or degree of tolerance for error. Within these limits, the tolerance is considered acceptable and the product will function as designed. However, outside these limits the product will not function satisfactorily and would be considered defective.

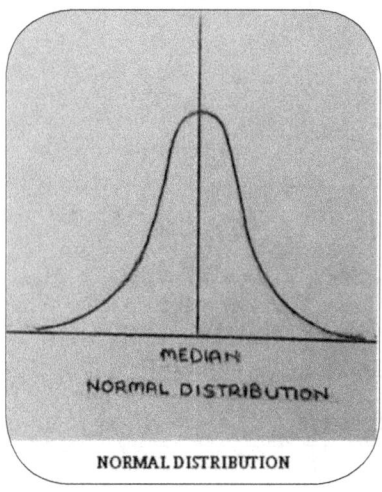

NORMAL DISTRIBUTION

A defect is an indication that something is wrong and needs to be corrected. Tolerance is allowed around a fixed reference (the drawing specifications). If we look at this mathematically, this is typically represented by a normal distribution, around the median or reference (see diagram above) for multiple samples of similar dimensions or characteristics. Most samples are close to the median in the form of a normal or equal distribution on either side of the reference. We can then look at the distribution data and calculate the standard deviation. Based on the acceptable tolerance, one standard deviation may be the acceptable tolerance on either side of the median. Any reading outside one standard deviation would be unacceptable.

The diagram below shows two standard deviations, one on either side of the median. This would represent the typically accepted tolerance in a manufacturing process. Any component within this range is acceptable and anything outside would be rejected.

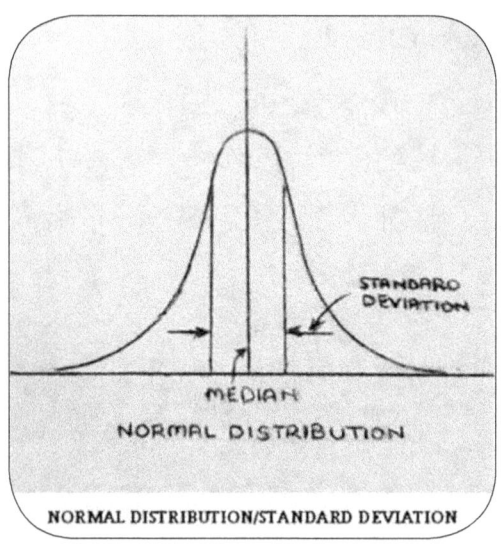

NORMAL DISTRIBUTION/STANDARD DEVIATION

An example of the application of tolerance is a piston that must fit into a cylinder. If the radius of the cylinder is six inches, the piston must be smaller in order to fit inside the cylinder. It must also have a certain amount of clearance so that it will move freely in the cylinder. We must remember that the cylinder also has a tolerance. A six inch cylinder would have a tolerance of say, plus or minus .005". Therefore, the smallest acceptablediameter would be 6" minus .005". The largest diameter of the piston must be this value (6 -.005)", less the required clearance between the piston and the cylinder, so that the piston will move freely inside the cylinder. The tolerance value must be fixed by the allowable gap between piston and cylinder so that the there is a good fit. If the fit is too loose, there will be too much play and if too close, the piston will not fit or be too tight. One can now see the importance of getting the right tolerances to make a successful product.

Mass Production

The process of mass production involves timing, sequencing and repetition. This is important when we need to make a product, at a high production rate, with a continuous production rate. To set up such a system, it is important that all stages of production be coordinated including raw material availability from suppliers, warehousing for raw materials and finished products, and layout of the manufacturing process as far as sequence and quality control are concerned. Typically, there are quality control procedures at each component production stage.

On any production line, there are sequential stages of production. If we look at the complete system, we need to take into consideration everything from design to completion and even use, as one will need to consider product safety and reliability based on the history of the product during use.

Parts made must come to the assembly line when needed for the assembly process. This has to be planned and monitored closely to ensure that the process progresses smoothly.

The process starts at the concept or idea for the design, with research and development and then manufacture, starting with the prototype. It is also necessary to have customer feedback so the necessary changes can be incorporated in future designs that will better satisfy consumer needs.

If we look at our natural world, we see that there is production or shall we say, reproduction of living things. Since the materials used all follow the same laws both in our production lines and in the processes we see in nature, do you think it is reasonable to assume that a similar process was used for the design and reproduction of living things? There had to have been an idea and a plan, the source of which must have been intelligent. In addition, there had to have been sequencing and timing which further complicates the production/reproduction process. Again, this can only be done by an intelligent mind. No ordered system can be created or produced without following these sequential and timely steps, as well as knowing when to stop.

In the natural process, our food is the raw material that builds and maintains us. All the necessary elements used in the design of our physical bodies are in the food that we eat and air we breathe. The DNA (also genes) does the design, manufacturing and quality control.

The initial design must have been done by an intelligent designer as was the sequential developmental process.

Robotics

Robots are designed to simulate human body motion. Those who have designed robots know that this is a very challenging prospect as human motion is very complex.

Let us look at a simple production line, where it is necessary for human motion to be simulated in areas such as filling and packaging. Where the human being is replaced by a robot or automatic process, everything has to be precisely located. There is very little flexibility whereas, with a human being, flexibility is almost limitless. The difference is that the human being knows what needs to be done and when faced with changing circumstances, outside the norm, is able to compensate.

The robot has to be taught every movement and is unable to do anything outside the scope of what it has been programmed to do.

If a bottle to be filled on the production line accidentally falls or for some reason becomes skewed, it will not be properly filled and the robot has to be programmed to stop, if this happens. The human being would immediately recognize the problem and correct it.

The robot is given a fixed reference from which to start and all movements are referenced from that point. If the reference is moved, it will not work properly. If the reference is unchanged, the coordinates will be accurately located, but if the bottle to be filled is not in the right position, the liquid will spill unless the robot and production line are programmed to stop when such a situation arises.

The robot is designed to do highly repetitive work with great speed and accuracy, but only if conditions remain consistent with those for which it has been programmed.

Note- Robots could be designed with cameras that provide feedback similar to what our eyes provide to us. This would be an improvement on the basic design. However, to replicate the reasoning ability and emotions of a human being is still out of reach in this technology.

Computer System Data Storage

Everyone knows that inherent in the design of computers is accuracy. How can they achieve such accuracy? First they have to be given a fixed reference.

A computer stores data in binary form to the base eight. This means that it only recognizes two conditions, 0 and 1(binary). These are used in various combinations using 8 pieces of descriptive content made up of this binary format. Eight was chosen because it allows a relatively infinite number of combinations for representing each symbol stored or processed. Symbols meaning numbers and letters of the alphabet, etc.

Each symbol is made up of eight bits using only 0's and 1's. This is how simple it is made. In electrical language, this means zero voltage (or current) or one unit of voltage (or current) the value of which has to remain constant for it to be accurately recognized. This may be represented by only a fraction of a volt so low voltages can be used as long as it is consistent and recognized as such by the computer circuit. It only has to recognize the difference between 0 and 1. In computer circuits, the power supply voltage has to be highly regulated and stable so that it provides a stable, accurate reference for the computer circuit voltage.

Each symbol has its unique combination incorporating the eight bits. Symbols representing our alphabet, our numerical system and other symbols such as commas, periods, colon and semicolon are all represented. When the eight bits in the combination representing a symbol are input, say to the memory, the circuit recognizes that the information representing that symbol is complete and the next eight will represent the next symbol. They are stored in units of eight. Each of the eight spaces is filled with zeros and ones in the specific combination representing each symbol. See example below.

00000001, 00000010, 00000011 represent 1, 2, and 3, respectively

There is a specific combination for letters and for numbers so each is readily recognized. Each number or letter has its own combination, starting from the right hand side of the array. Each symbol is therefore clearly represented leaving little or no room for error. The only problem would be if the computer mistook a zero for a one, or vice versa. Care is therefore taken to ensure the accuracy of the voltage (or current) used to represent 1 and 0 and to prevent interference from outside the system.

As long as these are consistent, accuracy will be assured.

We can therefore store a name or a number, a sentence or paragraph or a book using these combinations. The computer is very fast because computations are done at about 1/100 the speed of light, at the speed of electrons flowing in a conductor.

It is therefore clear that order is critical and to make this possible, it is necessary to have fixed references for each operation to proceed in an ordered manner.

Computer Data Entry/Processing

If you have worked with computers you know how precise you have to be in inputing data. This is an indication of the precision needed for the system to accurately interpret the input so that an equally accurate output can be guaranteed. By now you must have heard the term ' Garbage in, Garbage out'.

Do you think you could randomly program a computer? To the contrary, every bit of information has to be accurate and in the proper sequence. For the same reason, the process of evolution could not have programmed a human being. It had to have been an intelligent designer.

Every command needs to be precise and in the proper sequence for us to even hope for an accurate output. This is an indication of how the data input into nature had to be precise to ensure that life was made possible and even more so, the creation and development of a human being.

When you think about this, do you believe that life was created from a random data entry process or was it an ordered process by some super intelligent mind?

Semiconductor Integrated Circuits

Anyone familiar with the manufacturing process of a semiconductor knows that the process requires the ultimate in quality control. There can be no contamination of the 'wafer' except for the intentional introduction of additives that modify the characteristics of the finished product, based on its intended function. This extreme level of quality control is also necessary to ensure replication to very high tolerance or low error rate. The level of rejects can sometimes be extremely high in order to maintain the desired very high quality standards.

This is what it takes to make one of the critical components in a computer. This is the only way to ensure accuracy and reliability when the product is incorporated into a computer circuit. It is far from a random process. It is totally ordered.

Television Data Transformation

The image we see on our flat screen TV is transmitted in the form of a matrix or pattern using radio frequency waves, as the carrier, which are part of the electromagnetic spectrum. This frequency is longer than that of visible light.

In order for the picture to truly represent the image in front of the camera, when it gets to the viewer's end, the bits in the matrix have to be sent to the screen in accurate sequence. The output must truly represent the input. There is only one way to do this and that means accuracy down to the last bit. Otherwise, the picture we see will be a distortion of the original or there would be no picture at all. To ensure accuracy, there must be a reference that stabilizes the signal so that the sequence cannot change. To achieve this, each station transmits each signal at a specific frequency so when we tune to this frequency we can receive the signal. The picture and sound data are superimposed on the carrier frequency in the same sequence in which they are recorded and transmitted in the same order. The sound and video information are then separated in the receiver and sent to the respective circuits to amplify the picture and sound information before being sent to the picture monitorand the audio output sections, respectively. The bits in the matrix, as we see in the specifications for the TV monitor display, for example- '1080' per square inch, each has a specific place on the screen and there is no room for error. There is no room for randomness.

Our interpretation of the picture we see, as far as continuous motion is concerned, is dependent on what is called 'persistence of vision'. The same phenomenon applies in our interpretation of a motion picture. When we look at an object, the picture we see makes an imprint on the optic nerves (retina) but does not immediately disappear. There is a delay of the imprint on the retina. This delay is short but is enough for us to perceive the entire picture as well as continuous motion even though the picture on the TV screen is being sent as individual bits of information sent in rapid sequence. The bits are being sent at about 1/100 the speed of light (2,000 miles per second) and so there is a lot of information being sent, per unit time. As the picture changes, it appears continuously in motion even though it is a series of individual bits sent in rapid sequence.

The brain interprets the information transmitted from the eye in such a way that we see color, contrast, motion and distance or depth.

Do you have any idea of the degree of order it takes to develop the eye and the brain working together for us to have such vision? Do you think it was randomly achieved through a process of trial and error, without a fixed reference? Do you think such a system could be obtained without input from an intelligent designer?

If this is the type of accuracy it takes for success at the data transmission level, how much more accuracy would it require to develop a system as complex as the human body?

If we went to another planet and saw a television set in operation wouldn't we believe something intelligent must have made it? We can probably use anything as an example such as a car, a plane, a house, etc. Yet, some of us believe that the human body was made from randomness, as well as put together randomly, over a long period of time.

Would you describe the human body, with all its complexity, as a random or as an ordered system? If you say random then we would have to use another definition for random. If ordered, that would make more sense. Remember any ordered system has to have a fixed reference. This reference cannot change and must be of an intelligent designer. In the case of the television set, man is the intelligent designer.

Radio Transmission

Like video transmission, radio transmission also uses radio frequency waves which are part of the electromagnetic spectrum. These are longer than those used for television transmission. You must have heard of AM and FM. These are amplitude modulation and frequency modulation, respectively. AM uses the lower frequencies of the radio wave spectrum and FM the higher frequencies, for transmission.

Sound waveforms are superimposed on an electromagnetic carrier (radio wave), in one case modulating the amplitude and the other, modulating the frequency. The sound information can then be transmitted at the speed of light and decoded at the receiver to get the sound information only. This is done using electronic filters. After decoding, the sound wave, in the form of an electrical current, is then amplified and fed into a speaker which vibrates the air at the frequencies fed into it, thus reproducing the sound.

We now transmit in digital format but this format must accurately represent the analogue format and be converted back to analog as this is the only way we would be able to hear or interpret sound.

When you tune your radio, it must be to the electromagnetic carrier frequency you are trying to receive, otherwise you will not find it. The associated electronic circuits must be precisely tuned to that frequency. This is the reference frequency.

Construction

Starting from basics, if we want to construct a building, we must first decide on a unit of measurement. This may be the inch, foot, meter or whichever we choose to use. However, once we chose a unit of measurement, we must stick with it or use some accurate form of conversion. This is now our reference unit of measurement.

In that chosen system, for any dimension to be accurately represented, it must be described relative to the origin. The origin is therefore the fixed reference. This is necessary in order to communicate accurate dimensions on the system 'blueprint'. One must be able to accurately interpret the magnitude and direction from that reference point to any point on the blueprint.

(What is referred to as the nominal value on a blue print or other engineering application is one which is acceptable and within the desired tolerance.)

As in any three dimensional system, we need X, Y and Z axes for it to be accurately represented. The X and Y axes represent the plan view and the Z axis the elevation. Here again, we must choose a fixed reference, which is typically given the coordinates, 0,0,0. We originate from zero on each axis. Each axis represents the magnitude and direction in each dimension. The distance from the reference is the magnitude. We need to select this reference and this cannot change after we start. If it is changed, the representation will be in error.

So, where did the reference come from? It came from a source from outside the system, the designer. It takes an intelligent decision maker to determine what is required including the starting point (the fixed reference).

Here again, tolerance is important especially in the structural members of a building. The minimum strength on a structural member must be met for it to be acceptable. In other words, the lower end of the tolerance must be at or above the minimum strength requirement. The required load specification would also include any safety factor.

Architecture

This is a combination of art and engineering. In order to create a structure that is aesthetically pleasing and also functional, we combine the two disciplines, art and engineering.

Art is both an emotional and spiritual expression of ourselves. To express this in material form we must obey material laws.

A building is designed with functionality in mind. In some cases, the design is to the individual taste of the client. After the aesthetic design is agreed upon, the next step is to calculate the structural properties to ensure that the structure can be safely constructed for the intended use. This is the job of the structural engineer.

At this point, the calculations are done on the loading of the structure including the footing, floor, walls and roof- in the case of a building. These calculations take into account the forces that the building will be subjected to including a safety factor, which is an additional load factor over and above the basic design loads. These factors are calculated to withstand static and dynamic loads. This takes into account weight (force), and distance or length (moment). These elements are critical in the calculations for the design strength.

Again, we see that fixed references are critical. All the specifications involved are predetermined and the minimum/maximum quantities set or fixed. The building can now be safely constructed to these design specifications.

Art/Painting

In painting, the primary colors for pigments are magenta, cyan and yellow. In order to achieve the brightest color, one can mix no more than two of the primary colors. I typically use the most brilliant of these two primaries such as chromium yellow and red for a brilliant sunset.

I cannot use blue in the mixture otherwise the color will be muted or greyed and so less brilliant. By adding blue I will have added the third primary which grays or neutralizes the color. This is the reason why when we clean our brushes, in water or turpentine, the color of the solvent eventually becomes muted or dull. At some point in our painting, we will have introduced the three primaries to the mixture causing the color to become muted. You see, with pigments, the three primaries added together cancel out each other.

The other colors on the pallet are variations on these primary hues obtained by mixing the primaries and the various hues of natural or synthetic pigments.

One must follow a fixed protocol when mixing colors, from the primaries or other hues, to obtain the desired hue (color). Each protocol is consistent and repeatable. For shadows, we add the complement to the hue under consideration. For contrast of a given primary or hue, we juxtapose (place adjacent) the compliment. In other words, for yellow to look its brightest we juxtapose the compliment purple which is the combination of red and blue. This provides the most contrast based on how the eye perceives color. If we want orange to look it's brightest we place it beside blue. If we want to make a shadow, we add the compliment to the color. This grays or neutralizes it.

It is interesting to note that this is the case with pigments as used in painting but, with 'real light', when we mix all the colors of the spectrum (in phase), we get white light and not darkness. This is because pigments absorb the parts of the spectrum that they do not reflect. They appear as the color they reflect. If the pigment is orange, it absorbs blue light from the spectrum and reflects yellow and red. If we add blue it now absorbs the red and yellow as well and so will now appear gray as it has already absorbed the blue with the original pigment. All three primary colors are being absorbed. Hence the color now appears gray.

This is one of the basic principles one has to know when mixing colors for painting. These relationships never change and we depend on them being fixed otherwise we would not be able to mix colors and consistently achieve the same results.

This only involves the mixing of colors, which is a critical part of any painting process. Then we move on to composition, the art of placing the colors on the paper or canvas. This requires us to have a subject or if abstract, a clear idea of what we want to convey to the viewer. These are the fixed references. We use our intelligence to fix the references and proceed from there.

As we develop a painting, we must use these 'consistent' principles of color mixing and subject matter and refer to them continually as we progress. These are our references and will always influence the finished product.

As mentioned above, it is interesting to note that light and pigment display opposite properties. If we mix the components of light, the colors of the spectrum, they enhance each other. However, if we mix the primary pigments red, blue and yellow, they neutralize each other and we get grey (or black).

When painting, to obtain the desired color we use the primaries as a reference as they are unique in the fact that we cannot mix any other colors to obtain the purity or likeness to a primary color. However, we can use the primaries to obtain other colors.

White light is direct light from a source that includes all the colors of the spectrum in phase with each other; or similar light that is reflected. If a surface appears white, it has reflected all the light falling on it randomly, without absorbing any of the colors in the spectrum. If the surface is mirrored, it reflects all the light at the same angle as the angle of incidence. This is the same angle at which it contacted the surface. The surface will then appear like a mirror. Chrome plated surfaces have this mirrored property. A black surface absorbs all or most of the light.

The characteristics of the surface determines how the light is affected based on which colors are reflected or absorbed. The surface itself has no color, only the light that it reflects has color. This is why in a dark room everything is black. No light is reflected. Beauty is in the light as this is the only medium that our eyes perceive.

Color Interpretation

Different colors are interpreted differently by the brain. Reds, oranges and browns evoke warmth whereas blues and greens are cool and evoke peace and calm. This may be the reason why the sky is blue and the trees are green. We need these colors in abundance around us to help maintain calm in our lives. This is the natural response in the way the brain interprets these colors.

Art is personal although, to some extent, cultural in our reaction to any particular work of art. Through art, certain emotions are evoked in us such as peace, turmoil, love, hate, etc. These emotions are triggered by the scenery or work of art which is being viewed.

The reference for color is light and this is fixed. This is why I now believe that we all see colors the same way. The yellow one person sees is the same yellow another person sees. In some cases, the brain may interpret it differently because of distortion due to a defect in the interpretation of the signal. That is an abnormality such as color blindness. However, I believe most of us see the same yellow, red or blue hues.

We therefore conclude that light is the reference for the color and shape of an object. All the frequencies are fixed and never change. Light is the fixed reference.

Light connects the universe. It travels long distances at great speed and makes it possible to see other planets and stars. For stars, this is the light generated and for planets, the light reflected.

Music

Music has a common scale of notes where the reference is middle C. It is mechanical vibration or sound and will be described in a later chapter. Instruments such as the piano and guitar are tuned using middle C as the reference. It is important that instruments be in harmony especially in a large orchestra or band. Otherwise, we immediately notice the discord. Our brain is looking for harmony and it becomes very disturbing and annoying when we hear discord. We are inherently ordered and will readily recognize any form of disorder as this is unpleasant to us.

Using middle C as a reference (261.6 HZ), we can compose an infinite number of tunes and will never run out of combinations. We have developed a musical scale using octaves (8notes) as incremental bands, before repeating, both above and below the reference. Octaves are frequency bands which the ear interprets as eight separate notes.

Human hearing is within a specific frequency range from about 20 cycles per second, at the low end, to about 20,000 cycles per second, at the high end. This is for those of us who have excellent hearing. Hearing deteriorates with age and from consistent exposure to loud noises. Typically, hearing loss starts at the higher frequencies.

Our sense of hearing makes it possible for us to hear each other when we speak, to hear music, and to warn us of impending danger.

Our two ears, located on opposite sides of the head, allow us to locate the direction from which the sound originates. We not only hear sounds in stereo but we also have stereo vision. Stereo vision gives us the ability to judge distances. Do you think this all evolved randomly?

Sports

Every sport has to have rules. These rules define the sport and differentiates one from the other. The object is to achieve some goal whether as an individual or as a team. The goals are preset, with a scoring system to decide the winner. Typically, the individual or team with the highest score wins. However, in some sports, as in golf, the person with the lowest score wins.

The rules are known by all the participants and in some cases there is a referee or umpire who enforces the rules of the game. There are usually equal numbers of participants on the opposing sides. Also, the sport may be further refined by a process of elimination so that, in the end, the teams with the best record compete to determine the ultimate winner.

These rules of the game are references and they have to be fixed otherwise the game cannot be accurately defined and the final result will be in question. Each sport may be considered an 'ordered system' and so the essential components of an ordered system apply. The rules of the game are essential as they define the sport.

PHYSICS

THE LAWS OF physics define our physical or material world. I will now show you that these laws never change and work together to keep our universe ordered. These never changing laws indicate a fixed reference that never changes with time. Never-changing is one of the attributes of God. We see this fixed reference as it manifests itself in the following elements of our world.

Time

What do you understand to be the meaning of time? Is it the momentary display on the clock? Is it a finite period in which events take place? Is it the infinite period spanning billions of years? Is it a state of mind? I would say it is all of these things.

Time is the dimension in which the sequence of events is fixed. Some say time is relative, meaning it appears to go slowly or quickly, depending on our emotional state. If we are enjoying an experience, time appears to go quickly and if we are under stress, it can appear to go slowly, or even quickly if we are trying to meet a deadline. It all depends on the situation.

We use time as a reference for the sequence of events in our daily lives. If we did not have a memory, time would mean nothing to us. Time is recognized by comparing at least two events in our memory, one of which is fixed. It may then be said that, to each individual, time is a function of the mind giving us the ability to record events in the sequence in which they occur. However, time is real in that even if we are not aware, it is still progressing. Time progresses in one direction to maintain accurate sequencing.

At any given time, there are an infinite number of events occurring in our universe. In these instantaneous increments of time, which we refer to as the 'present', we are connected to each other. Once this instant has passed, the opportunity is lost until possibly some time in the future. What makes it even more difficult is that we can only be in one place at any given time. In other words, make the most of any opportunity as it may not present itself again in the future.

Our first memory is fixed at the beginning of what appears to be when life (time) began for us. When we are asleep or unconscious, as far as we are concerned, time stops. One may then say time is relative to each individual.

In our daily lives, time is our reference. Our current calendar is referenced from the approximate date when Jesus was born. Jesus must have made an indelible impression on the minds and lives of those who knew him, for us to now use the date of his birth as the fixed reference for time.

Time began at the absolute zero reference. In order for time to have been initiated it had to have had a fixed reference and still does. The reference cannot change because order has to be maintained. Time keeps all events in ordered sequence. The absolute reference, itself, is outside of the sequence of time because it initiated the sequence. Time is a product of this reference. Time is the dwelling place of the sequence of events that unfolds during our lives. The creator of time does not exist in this sequence. Thecreator is outside of time.

Time makes it possible for an intelligent mind to develop a plan, set a fixed reference, and initiate the sequence of actions necessary to complete that plan. It is the medium in which we are given access to event sequencing. We use time as a reference for the sequence of events in our daily lives. Again, we see that if we did not have a memory, time would mean nothing to us.

The only time we have available to us is the present. It is instantaneous, it is very short, yet everything we accomplish is in the present. The good thing is that the present is continuous and our accomplishments are cumulative. Yet the present is detached, in that, even though they are so close, we do not have physical access to the past or the future. If you want to maximize your accomplishments, you should focus on the present.

Age is a function of time. If time is not a factor then there is no aging. This concept is compatible with eternity. Time is relevant only to our realm.

The fact that every four years we add a day to the calendar is an indication of the approximation of our seconds, minutes and hours to the natural cycle. What we are doing is trying to maintain synchrony with nature's cycle. Nature is the reference and we have to synchronize with this fixed reference in order to maintain accuracy.

Time is designed to go in one direction because sequencing is critical. Order must be maintained.

Distance and Space

We live in a three dimensional world and so, in order to communicate location and relative separation, we need to define the concept of distance and direction.

In physics, this is referred to as a vector, having both magnitude and direction. We developed units of measurement such as the inch, foot, meter, mile etc. These are the standards or references for distance (magnitude). We have also developed 'degrees' to denote angle or direction from a fixed reference point.

In order for us to be consistent and accurate in relating or communicating distance and direction, we either have to use the same standard and use the proper conversion when changing from one standard to the other. It is critical that we have a standard or reference before we can proceed to using it in any practical application.

All we are trying to do is simulate and apply it to the three dimensional world in which we live, using a system to which all can relate. Using these standards, we have developed coordinates that are used to accurately define any point on earth or in the universe.

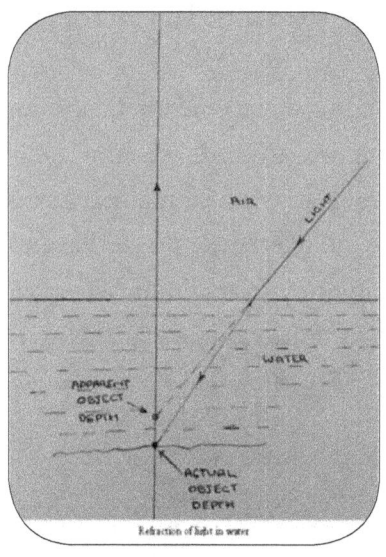

Refraction of light in water

We also use this system of measurement to define three dimensional objects and so it has universal application. It not only defines the size but also the shape of all things in the universe.

For infinite distances, we use light years which makes much more sense based on the size of the universe. For those who may not know, a light year is the 'distance' light travels in one year. At a speed of 186,000 miles per second, we can see why the speed of light is used in calculating the distance of stars or planets billions or trillions of miles from earth.

Light is a part of the spectrum of electromagnetic radiation, all of which travel at the same speed. They travel in straight lines in a vacuum. If there is no change in the medium in which they are traveling, they will continue in a straight line. In space, they travel in straight lines but variations in atmospheres do bend light to varying extents.

When light goes from air into water, refraction takes place, the light beam changes direction. This is why water appears shallower than it really is, when viewed from above. (See Fig. A next page) not 'below' as it is now.

Navigation

Now that the measurement standards have been developed for defining distance and direction, I will look at navigation.

Using the same standards of measurement used for distance and direction, with some modifications, they can be incorporated into navigation.

For navigating a three dimensional world, three axes must be used, the X, Y and Z, to represent the three dimensions. Now that there are three axes there has to be a set reference, a fixed reference.

For this purpose, we have used the equator (latitude) and Greenwich, England (longitude), as the X and Y axis (respectively), with the Z axis extending 90 degrees upward from the point of intersection of Greenwich and the equator (Fig. A next page). Using this point as a reference, any point on the surface of the earth can be defined as well as extended out from the surface, on the Z axis, into space. At the intersection of Greenwich and the equator the coordinates are (0,0,0), the fixed reference. The third 0 is the Z axis starting point. (Fig A next page) as in the one below.

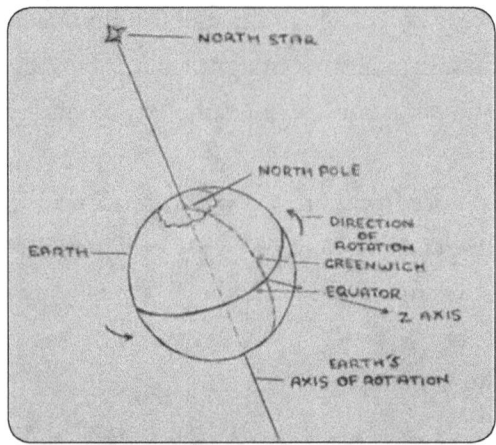

Fig. A Navigation Diagram

To define a point in space, there has to be at least two fixed points on earth and then triangulate to fix the third in space. In every case there has to be a fixed reference. All points are defined relative to the reference. (See Fig. A)

Before the current navigation system was developed, using strategically placed FIG. A. NAVIGATION DIAGRAM satellites in fixed orbits around the earth, the North Star was used as a reference. You see, the North Star is located in our galaxy at a point where it is on the same axis as that on which the earth rotates. It is located on the side of our North Pole, hence the name North Star.

If one were to draw a straight line from the North Star to earth, passing through the North Pole, it would be on the same axis as that on which the earth rotates. It was, therefore, the perfect reference for ships navigating at nights, since as far as the earth's rotation is concerned, the North Star is fixed. Satellites are now used in fixed orbits around the earth because we are able to strategically place them in orbit. By triangulation, any coordinate can be located on earth. If you know the coordinates, then your GPS can be used to take you to that coordinate anywhere on our planet.

Mechanical Energy/Kinetic Energy

One of the first equations one learns in physics is that the mechanical force on an object is defined as the mass multiplied by the acceleration of that object.

$$Force = mass \times acceleration$$
$$Also, Acceleration = velocity/time$$
$$Power = force \times velocity$$

Now, all matter has mass so every object has mass. If sufficient force is applied to an object, it will start moving and accelerate if the force is strong enough to overcome friction. In other words, if the mass is constant, as for any given object, the acceleration will be proportional to the force on the object. Acceleration is the rate of change of velocity.

This relationship was developed from experimental data. The above equations can be used in the estimation of the maximum speed of a car in relation to the power of the engine. It is universal in that it is applicable to any object being moved by a force of any origin.

These equations are used to simulate motion on earth. Simulating natural occurrences was the basis for their development so that we can predict how mechanical forces act on objects.

For objects moving under the force of gravity, the same principle is true. Gravity is a constant force acting on an object. It is generated by the mass of the earth resulting in a type of magnetic attraction on objects within its range.

This force has been found to be constant resulting in an acceleration of 32 feet/second/second, on objects close to its surface.

Technically, the force on an object only produces kinetic energy if the object starts to move. In this case we say that work is being done on the body which gives it kinetic energy. (Kinetic energy is energy due to the movement of an object)

Potential Energy

Potential energy is the energy in an object by virtue of its position. It is stored energy. Gravity is one force that gives an object potential energy. This stored energy can be released by letting go the object. Potential energy may be stored under controlled conditions, as in a clock spring or wind up toy. Under gravity, if you release an object that you are holding, it will fall to the ground. You can also jump from a high place and gravity will take you to the ground. You should always remember that, in order to gain potential energy, you have to do the work by climbing to an elevated place or lifting the object off the ground. To gain energy, you have to expend an equal amount of energy. You cannot get something for nothing. When you release the body the potential energy is now transformed into kinetic energy (by virtue of its motion). The different types of energy are therefore, interchangeable.

Electrical Energy

In circuit theory, electricity is defined in terms of current and voltage in a relationship that is proportional to the amount of resistance in the conductor in which the current is flowing. If the resistance remains constant, as in a copper wire of fixed length and diameter, current flowing in the wire will increase with voltage in a linear manner.

$$V = I \times R$$
$$V = \text{Voltage}$$
$$I = \text{Current}$$
$$R = \text{Resistance}$$

The resistance, R, is constant in any given conductor. The voltage or potential difference is used to drive the current, which is a flow of electrons.

Since electrons are negatively charged, they will flow towards the positive pole, as in your car battery or any battery for that matter. Electrons are always negatively charged. The term conductor means that the molecules of the material have free electrons that will flow through the conductor when a potential difference(voltage) is established. Current is defined as the flow of electrons. Electricity can be obtained from a chemical reaction as in a battery or from a conductor moving in a magnetic field, as in a generator.

The equations developed from these relationships never change. The relationship is true even in the most complex electrical circuit. If it were not so, we would not be able to pass the first step in building on our knowledge of electricity.

Having mastered the basics of electricity, scientists progressed to electronics involving vacuum tubes and now semiconductors. Here, they are able to control electrical currents and voltages and channel them in a desired direction. They can also accurately amplify electrical currents and voltage with a superimposed sound wave, as in the case of an audio amplifier.

In electronics, in addition to the basic resistance in a circuit, there is capacitance, inductance and impedance as capacitors and inductors are introduced into the circuit in order to modify its characteristics, based on the application. Impedance in an electronic circuit is analogous to resistance in a purely electrical circuit. These components have unique characteristics that make the circuit an 'electronic circuit', with the vacuum tubes or semiconductors, as opposed to a purely electrical circuit. Here, the electrons in the circuit are being manipulated to produce a desired output, based on the given input.

We now use integrated circuits to store information for artificial intelligence (computers), which we can access in microseconds in the proper circuitry.

Electricity is the perfect medium for information transfer in circuits involving artificial intelligence, just as electrical signals transfer information in our brain, using a chemical process. Information travels quickly, at about 1/100 the speed of light. Again, we depend on the fact that the laws governing electron flow or current flow are constant and hence predictable. In all these calculations we have a reference that is necessary in any ordered system. The relationship between the three components referred to (voltage, current and resistance) must remain constant.

When the brain is examined, to determine if it is functioning normally, we look at the electrical current activity to determine if it is within normal parameters. This is one of the parameters used to analyze brain function.

Circuit stability is critical in any given electrical or electronic circuit. This means that the relationship between the components in the circuit must remain constant. Otherwise, we would not be able to predict the outcome or, in this case, output. Here again there are tolerances on the values of the components in a particular circuit design. The closer the tolerance of the components to the design specifications, the more accurate will be the predicted output.

For these characteristics to remain constant, there must be some form of fixed reference. If we duplicate this design in another independent circuit, then we can expect to accurately duplicate the output. In electrical and electronic circuits, the line voltage is the reference. This is the main supply voltage to the circuit and must be kept constant. Variations in line voltage will affect the relationships in all other parts of the circuit. It is assumed that the characteristics of all components in the circuit remain within tolerance.

Only if physical laws remain constant can such predictability be guaranteed. They always do. Thus, once gained, we can build on this knowledge with confidence.

Light

Light is the part of the electromagnetic spectrum which is visible to the eye. Only a specific set of frequencies is in this range. These are from red, at the longest, to violet at the shortest wavelength, that are visible to the eye. The combined visible spectrum appears to us as white light. However, when broken down, there are seven separate frequencies between red and violet as seen through a prism or in a rainbow. It is this separation and infinite combination of these colors that allow us an unlimited pallet when we view an object.

White light is made up of fixed frequencies within the visible spectrum that manifest themselves as violet, indigo, blue, green, yellow, orange and red.

Objects that we see have the color or colors of the light that they reflect. This ensures that there is one reference (light) as opposed to each object contributing to its own, unique color, independent of this reference.

Color manifests itself only in the specific properties of the surface of an object. The red rose is designed to reflect red light and absorb the rest of the spectrum. The various shades of red are developed by the reflection of some of the adjacent frequencies such as orange and yellow. If white is included in the petal, this means that part of the petal reflects all light randomly and this is interpreted by the eye as white light.

Color is therefore a property of light and not of the object seen. The object itself has no color. Light is therefore the reference. There would be no color except for the existence of light.

We possess an innate appreciation for color to which we may react differently based on our individual taste. Typically, blue and green are considered cool and, red and orange warm. We therefore react subconsciously, in this manner, when presented with these colors as represented in our environment or in a work of art.

Color, as represented by hues and shadows, molds shapes which communicate the unique characteristics of what we see.

Light transmits all the information we need to detect and interpret what we see. Light itself does not add or subtract from what is communicated to us but is consistent and true in its properties. In this way, it accurately depicts the environment around us. It is used as the reference or standard in the electromagnetic spectrum which we routinely use as the medium with our sense of sight.

You have probably heard of infrared and ultraviolet radiation. These are as named, the part of the electromagnetic spectrum before red light and the part beyond violet, respectively. Infrared is a good heating source and is often used to cure paints more quickly and for tanning. Ultraviolet (black light) also has its uses as certain chemicals glow when exposed to ultraviolet and can be readily identified.

It is also useful as a disinfectant. Because of its higher frequency, it is more penetrating and can cause damage to the skin, if overexposed.

Light frequencies within the visible spectrum never change and so we have a fixed reference. If there is no light we are visually cut off from our surroundings.

X- Rays

X-rays, as well as gamma rays generated by radioactive isotopes, are also a part of the same electromagnetic spectrum of which light is a part, but these are of shorter wavelength.

X rays are of even shorter wavelength than ultraviolet and therefore have greater penetrating effect allowing it to pass through the human body. The density of the human body varies depending on the type of tissue. The more dense the tissue the more X-rays are absorbed. This is how we can develop a picture of the inside of the body by exposing a sensitive film to the x-radiation after it has passed through the body. There will be variations of gray on the exposed film depending on the density of the tissue through which the x-rays passes. The developed x-ray film shows the bone structure of a lighter shade than the fleshy areas because it absorbs more of the x-rays. It is, therefore, like a film negative with lighter areas representing denser tissue and dark representing the less dense tissue.

For shielding an x-ray room and also for isolating radioisotopes during storage, lead lined enclosures are used because lead is very dense and will significantly reduce the amount of radiation escaping.

All these types of radiation are in the same category (electromagnetic) but have different properties. Our eyes cannot see microwaves, radio waves, infrared, ultraviolet, x-rays or gamma rays and are only sensitive to the visible light frequencies. Vision is limited to the visible spectrum which has all the information we need to enjoy a very satisfying visual experience.

Do you think that the process of evolution limited our sight to light frequencies in this entire spectrum just by chance. Or was this by design, by an intelligent creator?

Microwaves are electromagnetic radiation with frequencies longer than infrared and are also damaging to human tissue. To protect us, our microwave ovens must be shielded while operating. We must also be shielded from inadvertent exposure to x-rays, while these rays are being generated, as we can only have limited exposure without long term damage to the human body. Yet the visible spectrum is relatively harmless. The frequencies we need in order to see are those that are the least harmful. These are located in the range between radio waves and x-rays. We need to be exposed to light so it would seem that by design, we would be protected from overexposure to light.

Remember, the frequencies above and below light frequencies are very harmful to us. Why not visible light?

Sound

Sound, like light, may be mathematically represented by a wave form. It is mechanical in nature in that it induces mechanical vibration in the medium that it travels. It is detected by our ears as sound although some low frequencies are felt as vibration. Sound waves are audible to the human ear between the approximate frequencies of 20 cycles per second and 20,000 cycles per second. Sound travels at about 760 miles per hour in air and so is much slower than light. This is why when we watch fireworks, we see the explosion and there is a delay before we hear the sound of the explosion.

Throughout this range of frequencies, each one is fixed by a reference which is the number of cycles per second. The volume of sound that we hear is a function of the amplitude of the vibration. This is the distance the waveform cycles from the reference in the wave form diagram. The reference has to be fixed for us to make sense of what is happening. It is typically represented by a straight line. (Shown in the figure below.)

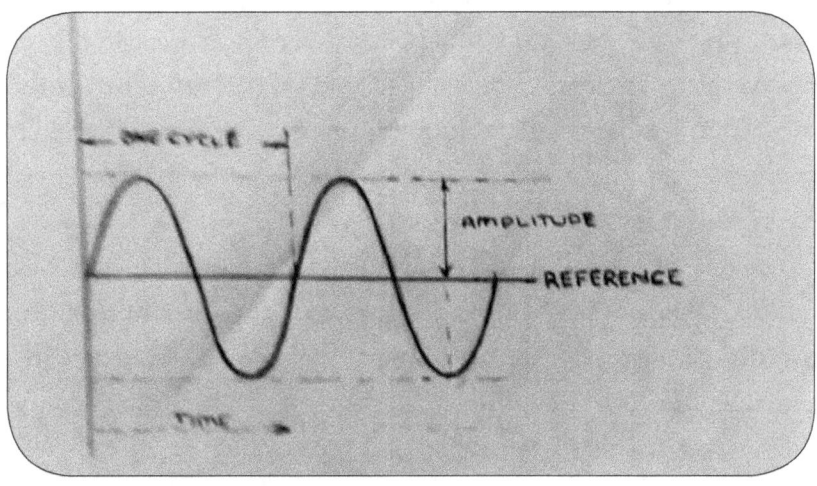

The Mathematical Representation Of A Waveform

Resonance

Resonance is associated with the transfer of sound (mechanical) energy to an object that vibrates in sympathy with the specific frequency called its resonant frequency. This varies from material to material based on its molecular structure. When the resonant frequency is reached, the material begins to vibrate uncontrollably and could even be destroyed. What is happening is that, at the frequency of resonance, the material will start to vibrate with increasing amplitude even if the sound energy exciting it is constant. This frequency is constant for a particular material of a specific length and thickness when, ideally, it is restricted at the ends. In other words, the material can be made to vibrate at its resonant frequency if the same conditions are repeated. The resonant frequency of a given material sample is constant and it will vibrate uncontrollably at that frequency.

This principle has various applications in real life such as in a horn, siren or even a switch. The human body has a natural resonance of about seven cycles per second. When subjected to this frequency, at an optimum magnitude, you will become disoriented and get sick to the stomach. This has been used for crowd control but is considered inhumane.

Sound makes it possible to verbally communicate with each other, to help us navigate our environment, as well as for pleasure in the form of music. Because each frequency has a fixed reference (the wavelength), we will always get the same result if the same conditions are repeated. It is consistent and predictable.

CHEMISTRY

Properties

Matter manifests itself in three distinct states- sold, liquid and gas. The properties of a chemical indicate the unique characteristics of that particular molecule or compound. Within fixed environmental conditions, these remain constant. We can depend on these properties to be consistent within each state. This is the reason why we are able to obtain consistent results in the reactions of the various chemicals. Without this consistency we would not be able to duplicate our discoveries in chemistry.

Periodic Table

We have found that, in nature, atoms/molecules manifest themselves in ascending levels of complexity. The Periodic Table begins with the hydrogen atom, consisting of one proton and one electron. Then, by adding another proton and electron we get helium. The series then progresses discretely with the addition of a proton and an electron to form new elements. These all occur naturally.

It is important that for every proton added there is a corresponding electron so that the molecule remains neutral and not positively or negatively charged. If a proton is added without a corresponding electron we now have an 'ion' which is basically unstable and it will try to find an electron to become neutral and stable.

Another particle found in the nucleus of an atom is a neutron. This is neutrally charged and is thought to be a type of glue that holds the 'all positive' nucleus together.

Valency

Atoms and molecules combine in discrete amounts to form compounds based on their valencies. Valencies are fixed and so an atom/molecule will consistently react with other atoms/molecules, based on these discrete combinations. Under the proper conditions, we can make chemical compounds with repeated success by understanding the conditions necessary for these reactions. The valency is a property that results in neutralization of the two reacting atoms/molecules which is in line with nature always trying to maintain stability. Valencies typically have values represented by integers such as 1, 2 and 3.

But, even in the proper valency proportions, the reaction will not take place unless the proper conditions are met. These are fixed conditions that must be met in every case. These conditions include temperature, pressure and environment. Environment could mean the introduction of a catalyst.

When we cook, we use fixed conditions and proportions to prepare food recipes even though we are only mixing compounds and may not be changing them chemically.

All atoms/molecules react in a manner based on their valencies, to form the compounds that there are today. We use these known principles to develop new compounds and are able to consistently replicate the chemical reactions.

This is so because the chemical reactions follow fixed laws that never change. We depend on this consistency and our ability to understand and replicate the necessary conditions for making the compounds we have formulated and continue to formulate.

In sequential order are all the known elements. At the lower level in the periodic table, the nucleus of these elements are stable but, as the nucleus increases in size, instability increases and they eventually become unstable or radioactive. This means they will spontaneously break down.

Atoms and molecules, in specific combinations, produce the various compounds that make up the earth and of which we are made. In living organisms, carbon is the base, with this group labeled organic compounds.

Starting from the elements of which living things are made and the inert elements around us, there is a consistency that is repeated throughout nature. We can therefore conclude that matter is made up of the same basic elements that exist in sequential order and form all compounds. This indicates that the basic building blocks of the universe are consistent, sequential in nature and ordered. There is, therefore, order in even which appears to be random at first glance.

Atoms and molecules combine to form the compounds that are the raw materials we use in manufacturing. In chemistry, we learn to modify or change these compounds through chemical reaction to form new compounds with new properties which we find useful in providing us a better quality of life-or so we believe.

These particles of matter and their ability to combine, all abide by specific laws that must be followed for these changes or reactions to occur. Under the same conditions, we can repeat these reactions to produce the same end products again and again.

Radioactivity

At the upper scale of the periodic table, the nucleus of the atom/ molecule begins to become unstable, hence the molecule becomes unstable, at the threshold where radioactive isotopes begin to consistently form. Some of these are highly unstable elements that generate harmful levels of radioactive energy (gamma radiation) in their natural state. Although radioactive isotopes have been used in medicine to treat cancer, they are inherently deadly to living tissue and must be used with great care.

The fission or breakdown of some isotopes releases intense heat which can be used for heating water to generate steam used to run turbines, to generate electricity and to power engines.

There are naturally occurring and artificially manufactured isotopes and have specific uses depending on their individual properties. Again, only certain elements are naturally radioactive and are consistent in their individual properties.

Laws of Nature

It is not important that you understand the mathematics of these laws as this exercise is only to demonstrate that they never change. For those who do not find this 'appealing', the only takeaway should be the fact that all laws governing material things are consistent and predictable. We cannot change these laws.

It is the inherent consistency of nature's laws which ensures that anything incorporating matter must conform to them and follow an ordered process, governed by these laws, to create and develop any ordered system. Randomness is not compatible with such a process.

The intent is to create order which can only be the product of an intelligent designer. Therefore, evolution, as we define it, cannot be the manifestation of such an ordered process, since, as defined, the process of evolution is not ordered but random.

If life processes were random, there would not be the interconnection and interdependence as we currently observe. We are all connected to each other and everything in the universe based on the fact that we all can interact with each other and our environment.

PHYSICS

Gravity

We are confident that when we drop an object it will fall to the ground. If we examine this further, we find that the object will accelerate, falling at a rate of thirty-two feet per second, per second (32 ft./sec./sec.). This is constant under gravity. The size, weight and density of the object are some of the variables. Yet, all objects will accelerate at the same rate if there are no other variables affecting them, such as wind resistance or some other external influence. The density, size and weight of the object do not affect this calculation.

A feather will float down slowly because of wind resistance and a rock will fall quickly. However, if we were to remove wind resistance by placing both objects in a vacuum and release them simultaneously, the feather and the rock would fall at the same rate of acceleration and hit the ground at the same time.

This experiment shows that the force of gravity is constant within the environment in which we live and has the same effect on all matter. However, in outer space, the gravitational force decreases until it becomes so small that objects float.

It has been found that the attraction due to gravity varies inversely as the square of the distance from the earth. When we are close to earth, the force of gravity may be considered constant and at a maximum but decreases exponentially as we move away from the earth's surface and into space.

There are many attributes of our universe that follow the inverse square law so we can use this principle in the formulation of the relevant mathematical equations.

Examples of these are the intensity of a magnetic field as we move away from the magnet or electrical source generating the magnetic field. This is also applicable to electromagnetic radiation such as light and x-rays as we move away from the source. This means that the intensity falls off exponentially, in proportion to the inverse square of the distance from the source.

The equation below demonstrates this principle. In mathematics, inverse means 1/ (the value under consideration). In other words, 1 divided by the value. In this case, it is the distance 'd' squared.

1/ (d x d) where 'd' is the distance from the source

As 'd' increases, the magnitude of the force begins to exponentially decrease. If the distance is 0 feet, the value will be the maximum it will ever be. If we measure this value 10 feet from the source and the force is F, when 20 feet from the source, which is twice the distance, the value will be 1/4 x F or 1/4 of what it was at 10 feet. Likewise, at 30 feet the value will be 1/9 x F. What we are doing is squaring the distance value thus resulting in an exponential decrease.

As far as the gravitational force of the earth is concerned, the equation is, the gravitational constant G x the mass of the earth(M) x the mass of the object(m), divided by the distance between the center of the earth and the center of the object(r), squared. The gravitational constant Gdoes not change, nor does the mass of the earth.

$$Force = GmM/r \times r$$

Since the radius of the earth is so large compared with any object on its surface r x r will not start to change significantly until the distance of the object from the surface of the earth is in the order of miles.

r = the distance between the centers of earth and the object. This would then be the radius of the earth + the distance of the center of the object from the earth's surface.

This equation was developed from experimental work done in the field of physics and found to be consistent. This can also be demonstrated mathematically since the force radiates in all directions and the intensity falls off as the square of the distance from the source. Again, this shows that nature displays consistency and predictability.

The inverse square law demonstrates that, close to the source, the effect of the force is at a maximum, but as distance from the source increases, at a certain point, the effect starts to rapidly fall off. This principle keeps the force, with its maximum effect, where it is designed to have influence and with relatively no effect at long distances from the origin.

In cases where the force is designed to have significant effect at long distances, the force is of an exponentially large magnitude. The sun, being a massive object, has a large gravitational force, enough to keep the planets in our solar system in orbit around it. However, it has no effect on the planets in other solar systems. Earth has an effect only on objects withinits gravitational field. Objects close to earth will be attracted and kept on its surface.

The moon is in orbit around the earth and maintains this orbit because there is a perfect balance between the gravitational pull of the earth and the equal and opposite centrifugal force of the moon in orbit.

In nature, where there is a law, there are no exceptions. These laws have been developed from close observation under controlled conditions in experiments by scientists. We are therefore able to build on this data and grow in our knowledge of our planet and the universe.

The reason for this exercise is to show you that the universe follows strict laws that do not change and that there are no exceptions to these laws. This indicates to me that the universe is an ordered system and could not have started from a random event but rather one which was planned and then carefully executed.

The laws, being constant, indicate that they have fixed references. If you observe nature, you will see that things only remain constant when they have fixed references. Otherwise, there is instability and chaos. The cycle of our day is 24 hours. This does not change because the earth takes 24 hours to make one rotation. Because this remains constant, we can use this as a reference for time. Earth also takes one year to orbit the sun and this also remains constant.

Here are some other examples of other laws that govern matter:

Matter Cannot Be Created or Destroyed

Matter is defined as anything that occupies space. When it is said that matter cannot be created or destroyed this means that the amount of matter that has always existed in the universe, since creation, is the same and will never change. It can, however, change from one form to another as in solid, liquid, gas and nuclear fission or fusion. The case of a nuclear reaction is where energy is released as the nucleus breaks down or nuclei combine, but the gross quantity of matter and energy remains constant in all cases.

As far as 'life' is concerned, there was one creation, but life is self-sustaining or self-perpetuating through the process of reproduction, manifesting itself in the form of living matter.

PHYSICS

Thermodynamics-Refrigeration

Physicists have discovered that a liquid must absorb heat from its environment in order to change from a liquid to a gas. This is called the latent heat of vaporization. During this transformation, the temperature of the liquid does not change. All the heat energy is used to perform the transformation.

It is also known that, if we compress certain gases, they become liquid at normal temperature. These are the typical refrigerants. Now, if the pressure is released, by forcing the liquid refrigerant into a larger space or volume, it will revert to gas as the pressure suddenly decreases (it boils). In so doing, it must absorb heat to make this transformation. The environment in which this transformation takes place is cooled, as a result of the heat absorbed by the refrigerant for the transformation. This would be the space in your refrigerator, or building interior in the case of the air conditioning unit. The evaporating refrigerant is passed through metal coils over which air is blown by a fan. The air is then cooled by the evaporating refrigerant in the coils and is used to cool the space.

The law of nature governing the evaporation of a liquid must be upheld and so the necessary heat to transform the liquid to a gas must be obtained from somewhere, in this case, the environment.

Using this principle, we make air conditioners and refrigerators or any appliance that is designed for cooling. The law is fixed and so we know we will get the same result every time.

Refrigerants have inherent properties that lend themselves to such applications.

The above characteristics are constant for any given system so we are able to duplicate any given system as long as we follow these principles. Again, this displays a consistency in the laws that govern matter.

Action and Reaction are Equal and Opposite

This is one of Sir. Isaac Newton's laws. It means that if we apply a force to a body it reacts with an equal and opposite force.

We use this basic concept to represent the dynamic systems that we routinely replicate in our inventions. An example of this is a jet engine or rocket engine. In the jet engine, air is pulled in, compressed, expanded and then ejected at the rear. The initial suction(action) of the air results in a forward force(reaction) on the airplane which is further enhanced by the thrust exerted by the compressed, heated air being ejected from the rear. In the case of the rocket engine, the burning fuel being ejected(action) creates a reactive force that propels the rocket upward.

These unchanging laws appear to indicate that they originated from an ordered, as opposed to a random source.

Electromagnetism

Electromagnetism implies a relationship between electricity and magnetism. Even though they manifest themselves differently and represent different forms of energy, they are interchangeable. An electrical generator demonstrates this principle.

If an electrical conductor is placed in a magnetic field and moved within the field, an electric current will be generated in the conductor. Conversely, if an electric current is run through a conductor, a magnetic field is generated. A generator produces an electric current by rotating a conductor (coil) in a magnetic field. The conductor is coiled to maximize the effect in that, as for all practical purposes, there are now multiple conductors. The effect is additive.

X-rays, a type of electromagnetic radiation, are generated by high voltage electricity in a vacuum tube, accelerating the electrons, with enough energy to generate x-rays when they collide with a fixed metal target.

Like light, these are also photons and a part of the electromagnetic spectrum. Light is the part of the spectrum with which we are most familiar. X-rays have a fixed frequency range (shorter than light) and are defined by their frequency range. The electromagnetic spectrum starts from radio waves at the longest to gamma rays at the shortest wavelength.

CHEMISTRY

The Ideal Gas Law/Boyles law

The ideal gas law enables us to develop a mathematical relationship between temperature, volume and pressure in an isolated gaseous system. This principle is used to develop the equation $P \times V/T = K$, is a constant. The symbols represent pressure, volume and temperature. In the equation below, the suffix (1) represents the initial conditions and the suffix (2), the final conditions.

Since $P \times V/T = K$, where K is constant, then it follows that, $P1 \times V1/T1 = P2 \times V2/T2$

If we keep one of these variables constant, we can define the relationship of the other two in the equations listed below:

with V kept constant: $P1/T1 = P2/T2$, since $V1 = V2$, then $P2 = P1 \times T2/T1$

with Pkept constant: $V1/T1 = V2/T2$, since $P1 = P2$, then $V2 = V1 \times T2/T1$

With T kept constant: $P1 \times V1 = P2 \times V2$, since $T1 = T2$, then $P2 = P1 \times V1/V2$

We can make any one the subject of the formula and so if we know the value of three of the variables, we can calculate the fourth.

If we increase the temperature in an isolated system (V is kept constant), in order for this relationship to be true, the pressure must also proportionally increase. This has been found to be experimentally true and verifies the unique relationship between these elements in an isolated system.

As indicated above, the suffix (1) represents the initial values of the elements in the equation and the suffix (2) represents the final values. This means that within a given isolated system, if we change one of the parameters, for example the temperature, with the volume remaining constant, the pressure will proportionally change. If the temperature is increased, then we would find that the pressure would proportionally increase.

Decreasing the temperature would decrease the pressure, proportionally. The relationship in an isolated system must remain constant. One of the variables is typically fixed, in this case volume, to make the calculation simpler.

$$(V1 = V2)$$

This is the basic principle used to design a pressure cooker. The lid is secured to the cooker so that the volume remains constant. When we place the pressure cooker on the stove and turn the burner on, the temperature of the contents in the cooker begins to increase. Since the volume is constant, the pressure will proportionally increase and will continue to increase as the temperature increases. The cooker is equipped with a safety relief valve to prevent it from exploding from over-pressure. P1 would be the initial pressure which would be atmospheric. T1 would be the ambient temperature. T2 would be the final temperature, that reached inside the cooker from the burner on the stove. Since the volume is constant and also known, the value of P2 can be calculated as this would be the only unknown.

This principle is also used in determining the potential power of an internal combustion engine. The volume of the cylinder remains constant. The volume above the piston does, however, vary from a minimum to a maximum, as the piston reciprocates. The volume used in the calculation would be the minimum volume at the top of the stroke of the piston at the time of ignition (this volume is constant).

With the volume at a minimum, at the top of the stroke, the fuel is ignited. This minimum volume is constant and a function of the size of the cylinder. The ignition results in a rapid increase in pressure on the piston, from the sudden increase in temperature, resulting in rapid expansion of the air/fuel mixture inside. The piston is forced down, this force is transferred to the crank shaft causing it to rotate. The cycle is repeated, resulting in the crankshaft continuously rotating.

If the temperature is increased in a confined space, the pressure will increase, based on this law. At the time of ignition, the volume is at its minimum. The higher the temperature of the explosion, the greater the pressure or turning force on the crankshaft.

In physics, pressure is defined as force per unit area and force is mass multiplied by acceleration. The larger the cylinder the greater the potential force. This force is multiplied by the number of cylinders and made continuous with the continuous cycles of the engine ignition system. The engine therefore becomes the driving force of the vehicle.

$$P = F/area, \ F = P \ x \ area \ P = pressure, \ F = force$$

This relationship indicates that the larger the cylinder or area, the greater the potential force.

Consistency of Laws

We depend on the above laws to remain true. These laws are the fixed references on which all our scientific knowledge is based. A fixed reference is an indication that there is continuous, consistent control over the elements of matter. This implies order, and order implies intelligence.

In order for us to try to deduce where the universe all started, we can only use the tools and the knowledge we have available. In other words, study the dynamics of the universe and look for patterns in its characteristics that give clues as to its origin.

These laws are just an example of the consistency with which nature operates. We depend on these consistencies as references in all things scientific. If they were continually changing they could not be used as references. We could make no progress in science.

The laws of nature are designed to have a specific ordering effect on matter and so there is a relationship between them that, combined, maintains order in the universe.

Thresholds of Matter

Thresholds mark the end of one phase and the beginning of another. A dramatic change in properties as it relates to matter. Where the laws of matter are concerned, these still apply, but matter now takes on new properties that will remain consistent above the new threshold, in the new phase.

Thresholds are apparent at the freezing and boiling points of molecules. The most familiar to us is water as it changes from ice to water and from water to steam or vapor. Ice has different properties to those of water and water has different properties to those of steam or water vapor.

It is a characteristic of the molecules of all compounds and is a function of temperature, volume and pressure.

Under any given conditions, we have found that these bear a constant relationship. Variations in environmental conditions only change the threshold (temperature/pressure) at which these changes occur.

An example of this is water changing from liquid to gas at its boiling point of 100 degrees Centigrade, at a pressure of 14.6 pounds per square inch (atmospheric pressure). This is what is referred to as normal pressure, at sea level. However, if we go up into the mountains where the atmospheric pressure is less, the boiling point is less than 100 degrees Centigrade because there is less atmospheric pressure to keep the water from boiling.

In each of the three states, solid, liquid and gas, the molecules exhibit completely different properties, but within each state the properties are consistent and predictable. This all adds to the diversity of nature but shows overall consistency within the confines of the laws governing them.

Another example of this is the threshold at which a nuclear chain reaction occurs, as in an atom bomb. Below this nuclear threshold, the molecules of the radioactive chemicals are 'relatively' stable but still highly radioactive. However, once the threshold is met, a chain reaction begins, resulting in nuclear fission with an exponential release of energy until the reaction is complete. To initiate the chain reaction, there is a detonator which must be activated in order to get the radioactive atoms to this chain reaction threshold. We can control activation by controlling the detonator. The reacting radioactive atoms must also have a minimum threshold of purity for the reaction to be successful. Once the nuclear thresholds are met, a chain reaction will occur after detonation.

Summary

The above are only some of the more popular examples of the basic applications of 'consistency' as it relates to ordered systems. It applies to all aspects of our conscious and subconscious lives which would not be possible without these 'TRUTHS'. This 'ORDER' is what connects us to each other and to the universe. Otherwise, we would be disconnected and unable to interact with each other and the rest of the world. In fact, we would not be aware of our existence but for the fact that there is 'ONE' fixed reference from which 'EVERYTHING' radiates and which connects us all.

Intelligence plays the most important role. It allows us to connect increments of knowledge and awareness to form patterns which we then use to come to the realization that we do exist and are a part of a greater existence. This intelligence and knowledge can then also be used by us to initiate change in our environment.

When we observe nature and look at the abundance of compatible life in one place, it would seem reasonable to conclude that it is by design. Such order cannot be randomly achieved.

Many of us believe that we are so smart, making discoveries and inventions- we should take pause and humble ourselves. We cannot create anything new, in the absolute sense. This knowledge was already in existence in another realm. The knowledge is only now being shared with you. The blessed ones are only the media through which these revaluations are transmitted. It is all already in existence in another realm even if we do not yet know it.

I will now share with you my conversion experience and try to describe GOD in the way he describes and reveals himself to us in the HOLY BIBLE. We will also look at what he has revealed about creation. I think what you will find is that creation, as we observe it, in every detail, is exactly the way he describes it. It shows a close connection between GOD and the fabric of creation, so much so that it leaves no doubt that he is the creator.

My Conversion

At an early age, I began to wonder about creation and evolution, as to which was really true. I thought there may be a creator but he/

she seemed far removed from me and out of reach. How did this all come about? How did I come to be here? I had nothing to do with it and that was obvious. I could see, hear, touch, smell and taste, the senses that made me totally aware of my being and existence. This was the reason I became so interested in science and, in particular, physics which describes the physical characteristics of the universe and the laws that govern matter.

I had always believed that evolution was the way living things change from one form to a more advanced form, although there were some things that could not be explained by it.

I was now 51 years old and working for an insurance company as a Risk Control Consultant. Although this was not my passion, I was grateful to have a job that could pay the bills and provide for my family.

It was January 18,1999, a Monday morning. I went down town Phoenix, AZ, to review the operations at one of our client's properties. This was a routine survey where I would report back to the property owner my findings and recommendations associated with the visit.

It was a high rise building and it would probably take about two hours to complete. As I went around reviewing documentation and completing a physical survey, my contact brought up the subject of God. I was not interested! I listened and answered questions in a polite manner but, in fact, I was looking forward to completing the task at hand and leaving as some of his questions and comments made me feel uneasy.

Towards the end of the visit, we were in the lobby, as I had finished the survey and was about to leave. Then, he said something that caught my attention. He said, 'I think you are close' meaning close to understanding and believing in God. As far as I was concerned that was so far from the truth. Then he said, 'All you have to do is humble yourself before him and ask for forgiveness and he will forgive you and reveal himself to you'. I had no idea what that meant but I said to myself it wouldn't hurt to try what he said if there was so much to gain, based on what he had been telling me.

I then looked away from him and, in my mind, said, 'If you are there, please show me how to live my life'. I said it with sincerity but was not expecting a response.

I then went to my car in the attached garage and for some reason tuned to the Christian radio station he had mentioned. He had also given me the address of the church he attended but I had no intention of going.

Throughout the rest of the week I kept tuned to the Christian station not even realizing that I was ignoring my favorite popular station. I still thought nothing of it but, in addition to listening to this station, I began to have a new appreciation for nature. Whereas before I believed in evolution, now I was believing in creation. I did not make this change, it was made for me! It began to get scary as I now realized something was changing me, I was not doing it and had no control over it.

I did not even remember that I had prayed for guidance but, as I looked back over the week's activities, I remembered my interaction with my contact that Monday morning and the prayer that I said to a GOD I did not know.

It was then that I realized that my prayer was being answered.

A real supernatural experience came the following Sunday morning, when I was in bed about to have a late morning 'sleep in' as I always did on a Sunday morning. I began to feel very uneasy but continued to try to fall asleep for a final nap. I had a sense of something or someone communicating with me repeating 'you should go to this church'. It was not in English, my only language, but it was clear what was expected of me. However, I still resisted. The feeling became so intense that I knew I had to go to this church. I knew I had no choice and if I disobeyed I would regret it for the rest of my life.

I had not been to church regularly for several years but today I had to go because some force, now inside of me and much more powerful, wanted this of me.

I got out of bed and told my wife I was going to church. She looked at me in shock and asked 'Are you ill?'. I said, 'no, I just have to go'.

I then called the church from the invitation information my contact had given me to get the time of the service and then got ready to leave. I got into my van and started on my way to the church. I had no idea what to expect but I felt something 'real' as I drove to the church. I felt peace and joy that brought a smile to my face. I do not remember what the sermon was about that day but I did know that I would be going back every Sunday.

This was 20 years ago and I still regularly attend this church and I am now a confirmed member.

The days and weeks following my conversion, I spent trying to understand what was happening to me. I would wake up at nights thinking it was all a dream but then realize that it was all real and I was being influenced by a source that was not of me.

I would now look at a leaf and see new beauty and complexity and say in my mind 'You made this!'. I would look at a baby and say the same thing. I now had an overwhelming appreciation for creation and the fact that I was a part of this great work and had a purpose in this life.

Would you believe me if I told you that I still needed more confirmation that GOD really existed and was communicating with me. The whole experience was still unreal but every time I challenged its reality he showed me that what I was experiencing was real.

Whenever I was confronted with his response to what I was experiencing, I was so overcome that I broke down in tears.

It was now Easter and the main focus in the Christian world is the crucifixion when JESUS died on the cross for our sins. I was on my way to a Walgreens pharmacy to pick up a prescription and listening to the radio to a pastor describing the events at the cross, at the time JESUS was being crucified. When he got to the part when JESUS was dying and said 'Father forgive them for they know not what they do'. The scene was so vivid to me that I broke down in tears and said 'Why are they doing this to you'. By this time I was sitting in the car in the parking lot and so had to wait several minuets to compose myself before going into the drug store.

I have never questioned GOD since that experience. I now believe in him completely.

Since my conversion experience I have never felt alone. I am always aware of his presence in me. It is a very comforting feeling, a feeling that all should experience. It is a feeling of peace.

When I described my experience to people that I met, I initially thought they would immediately become believers. To my surprise, this was not the case. In some cases I got this 'deer in the headlights look'. In other cases I could see them trying to understand but were unable to fully appreciate what I was trying to tell them. However, there was an immediate understanding by those who were already believers.

One thing that was revealing is that, looking back, it seems as if I was being prepared for conversion. I now know that GOD first humbles you and there is no doubt that he did that to me pre-conversion. There were crises in my life that I had no idea how to deal with and how to overcome them.

Conversion Final Thoughts

After someone is converted, one of the things that begins to happen is that an internal battle begins between your old self and the new. A transformation process begins. You instinctually want to maintain control of your own life. But, then you begin to see your flaws more clearly and that you have such a long way to go to be that perfect human being.

You become much more aware of yourself, your flaws and how you now need to act as a new human being, in GOD. This means character changes that you now have to consciously make in your new life. You are given a new heart. Your conscience is renewed. You want to be a good person and sometimes fall short, but you know the road you now need to travel. You get back on that narrow and difficult road and keep on moving ahead.

When you consider the alternative, you are reminded that this is something that you have to do even unto death. You have no other choice. Who else can you lean on and be confident that you can depend on in any situation- 'ONLY GOD'.

INTRODUCTION TO THE GOD HEAD

WE HAVE LOOKED at examples of the necessary components it takes to create and maintain an ordered system. The key components are intelligence and a fixed reference from which to build the ordered system. We also need knowledge but, for us, this grows with time as we observe, investigate and research the function of the planet we call home.

Knowledge is now being gained at an exponential rate. We are getting better at everything we do, learning how to control our environment to enjoy a better quality of life.

These key, critical components of order must have existed before we came into existence as we had nothing to do with the order already in the universe. The only explanation is that intelligence existed before the universe came into being and is responsible for initiating and developing what we are and see today. I say this because the universe is an ordered system and so its origin must also have been ordered.

So intelligence existed. Since we were not there, there is no way we can prove how life and the universe. However, we were given clues, written in what we refer to as the 'HOLY BIBLE'. Now, we can compare what we see around us to what is said in the BIBLE about GOD and creation, to see if they are consistent.

GOD tells us that one of HIS attributes is that HE does not change. HE also says that HE created the universe and everything in it. The latter would take infinite intelligence. So, we now have an intelligent source and a fixed reference, the critical components needed to create an ordered system. The ordered system is the universe and life as we know it.

He also tells us that he made us in his own image. He gave us intelligence and in so doing the ability to create. All we need is to have an idea, set a reference and start creating. We see that we do have these attributes, the ability to create, using the material available to us.

When we examine the evidence by observing what we see today and the order of it all, there is no other rational explanation. His fingerprint is all over the universe. The never changing reference and the infinite intelligence are reflected in all ordered systems that we see today. Everything we make and every positive act we perform have these basic components.

We do not have the ability to find GOD and HE only reveals to us what HE desires. But, what is true is that HE has revealed to us all we need to know in the HOLY BIBLE.

GOD gave us a jump start on the creative aspect of our attributes by giving us all the raw materials we need. All those atoms and molecules set out in an ordered periodic table. HE also made the laws to which everything material must conform and maintain constant control.

Now, we will take a closer look at how GOD describes HIMSELF and how this is reflected in everything we see in the universe.

God The Father

Do you believe in GOD? Do you believe there is some source 'up there' that directs what happens in this world or do you think it just happens randomly? Is it that same GOD that created the universe and everything in it, including life?

There is probably some reason for one to have any of these beliefs. However, if you think about it, you should come to some logical conclusion as to what you believe is really 'TRUE'.

Let us think objectively and see if we can review this together. Keeping in mind what we have already discussed on the consistencies in nature and the universe, we will now look for the similarities between what GOD says about himself and what we observe in the things around us.

God Does Not Change

One important attribute of GOD is that HE does not change. This is how HE describes himself in the Bible as revealed to us through the prophets with whom he communicated. This is also the most important characteristic of a fixed reference, a critical component of any ordered system.

> Bible references- (GOD does not change-Malachi 3:6, Hebrews 13:8 Jesus)

We have seen, in previous chapters, how critical it is to have a fixed reference when defining an ordered system (creation). GOD chose to reveal this attribute of HIMSELF to us because HE thought it was important for us to know this fact about HIM. Nature's (material) laws are constant and this indicates a fixed reference as their origin.

God Is A God OF Order

We see order in the universe and the planet on which we live. The reason for any disorder we see around us has been explained by what GOD refers to as SIN.

Order, in both the spiritual and material sense, means that all the laws governing both realms must be obeyed. Man was perfectly made and would have continued to be perfect if he had obeyed all of GOD'S spiritual laws. GOD gave man one command, that he should not eat from 'The Tree of Knowledge of Good and Evil'. MAN was even warned that, if he did, he would suffer certain death (eternal separation from GOD). MAN disobeyed this command and so had to suffer the consequence of being separated from GOD.

But even after this, GOD still had mercy on us and gave us a way to redeem ourselves. HE gave us HIS 'SPIRITUAL LAWS', in the form of 'THE COMMANDMENTS', to help us in this effort. But, because of SIN, we find it difficult, if not impossible, to keep these commandments. This is the reason, as we are, none of us is acceptable to HIM. We all break these commandments every day (SIN).

> Bible quote- (The Ten Commandments - Exodus 31:18, No one of us is without SIN Romans 3:10) (All have sinned - Romans 3:23)

It is interesting to note that we can break SPIRITUAL LAWS, because we were given free choice (free will), but we cannot break MATERIAL LAWS. SPIRITUAL LAWS are based on ethical references, which, because of SIN, we are no longer connected to the TRUE reference (GOD) and therefore unable to keep from SINNING.

There are fixed laws that govern the universe (MATERIAL and SPIRITUAL). GOD is a GOD of laws and will tolerate no dissent. A perfect system can have no flaws. HIS SPIRITUAL LAWS must not be broken and anyone who breaks them can no longer be a part of HIS spiritual order. You see, SIN is systemic and has affected everyone and everything since it was first committed.

God Created The Universe And Everything In It

GOD told us that HE created the universe and everything in it. When we look at the universe, it is ordered, life is ordered. Wouldn't it be rational to conclude that such order was a creation of a GOD who does not change, a GOD of order and of infinite intelligence? HIS intelligence is reflected in the complexity of creation, including life itself. We know that we are intelligent beings and we had nothing to do with the creation of our intelligence. Intelligence, therefore, must have come from our CREATOR. We have seen HIS fingerprint all over creation (it is ordered) with the interconnection of all systems and subsystems to a single fixed reference (GOD).

Bible reference - (God created the earth and everything on earth - Genesis 1)

God Makes All The Plans And Decisions Regarding Heaven And Earth

GOD planned creation and is in the process of carrying out HIS plan to the last detail. So much so that there is a set sequence in which events have been planned to occur and each sequenced event has to be fulfilled before subsequent ones can be initiated.

One important example of this is that the HOLY SPIRIT would only have been sent to us after JESUS had died for our SINS and returned to the FATHER. JESUS, himself told this to his disciples. These events unfolded exactly as planned (Bible quote - I will send the Holy Spirit after I leave -John 16:7). There are also several references in the BIBLE where JESUS fulfilled the prophecies in the old testament so that HIS actions would be as was written by the prophets.

When we read the Bible, we will soon understand that GOD, the FATHER, alone plans and makes decisions about everything in HEAVEN and on EARTH. Here, we now begin to see more similarities or the fingerprint of GOD in our universe. The universe came from a single source. HE tells us HE is the only GOD and there is no other. We also know that any ordered system not only has a fixed reference but, where there are subsystems, they are all controlled by the same, single source. In the human body, it is the brain that controls all systems. For heaven and the universe it is 'GOD the FATHER' who represents the 'BRAIN'. Even GOD the SON and GOD the HOLY SPIRIT do the will of the FATHER. GOD the FATHER is the planner and the decision maker.

Bible reference - (JESUS promises to spend the HOLY SPIRIT after HE returns to the FATHER - John 15:26)

Bible reference - ("I am your GOD and there is no other GOD but me")

Bible reference - (Jesus does the will of the FATHER -
John 4:34, 5:19-21, 5:30, 6:38)

We now begin to see the recurring theme or similarities between what we see in the universe and what GOD has told us about HIMSELF. Wouldn't it be reasonable to conclude that HE is telling the TRUTH?

Remember, HE told us that he created the universe. If that is the case then HE must have infinite intelligence. (Genesis 1)

Based on the above reasoning, GOD qualifies in every respect as being the creator of the universe. HE is infinitely intelligent, HE does not change (the fixed reference), HE does all the planning and is in control of everything.

God Made Us In His Own Image

The analogy between GOD and man is that GOD made man in HIS own image. Human beings are intelligent, with creative abilities. We plan our lives and can choose what we desire. We interact with and change our environment to suit our needs. Of all living things on earth, we are the most intelligent. Man was given authority over the things of the earth.

Bible reference - GOD made man in his own image
(Genesis 1:26)

God Is Eternal

God says 'I am the ALPHA AND THE OMEGA'. This mens HE is the 'BEGINNING' and the 'END'. In other words, HE is eternal.

Any ordered system that has a beginning must have an external initiator. To initiate an existence such as ours, there had to be an external initiator. This is the only way to explain our existence since we had a beginning and we had no input into the process. The only source that can be the absolute initiator is something eternal. It just exists, it does not need an initiator. It is ETERNAL.

> Bible quote (I am the Alpha and the Omega - Revelation 21:6, 22:13)

God Loves Us

The most important thing to note is that GOD says HE loves us. But we should also remember that HE is a 'JUST' GOD. He loves us but HE also has some expectations of us. HE gave us commandments to guide us as to HIS expectations of us. All expectations must be fulfilled. This is how SPIRITUAL order is maintained.

> Bible quote - (GOD Loves us- 1 John 4:16, John 3:)

The first and most important commandment addresses love. The love we should have for GOD and the love we should have for one another. Now, do you think that this is good or bad advice as to how we should behave? Do you not think it is better to show love to one another rather than have apathy or hate? Love is order and hate is disorder. Apathy has no particular bias just like randomness. The fact that GOD loves us is an indication that GOD is on the side of order. HE is order. HE defines order.

With the above reasoning in mind, it would seem to indicate that God is the true source of all knowledge and intelligence. If the knowledge we possess comes from any other source it cannot be trusted. It cannot be TRUE. There is only one TRUTH. GOD always tells the TRUTH. GOD is TRUTH. This is part of HIS character and one of HIS attributes. Whenever JESUS was about to say something profound HE always prefaced it by saying, 'Truly, truly I say to you'. The words that followed would be the absolute, universal 'TRUTH'.

> Bible quote - (Truly, truly I say to you - John 6:47, John 5:24-25)

> Bible reference - (GOD is a GOD of TRUTH - Isaiah 65:16)

SIN is what we commit when we disobey GOD'S will or the guidelines that HE gave us (the commandments). SIN, as HE describes it, 'separates us from HIM'. HE gave us the commandments so we would know what HE expects of us. However, we are unable to keep these commandments because we are now all flawed. SIN makes us flawed. We cannot help but commit SIN since we do not have the power to stop ourselves from SINNING. This is because we have lost our true reference. GOD is the true, absolute reference. Without his guidance, we set our own reference, which may not be in harmony with the absolute reference.

GOD is the only one who can give us power over SIN. At the time of your conversion, HE sends the HOLY SPIRIT to dwell within you and guide you. But even then you still SIN. The reason you still SIN is that you still possess part of your old self. Now, there is a continual battle between the old and the new self. Sometimes the old self wins, but ultimately the HOLY SPIRIT will prevail. This is what happens when you are saved and become BORN AGAIN, of the SPIRIT.

GOD is the one that calls you to HIMSELF and it is HIS will, as part of HIS plan. Some believe that HE calls us and also gives us the will to submit to HIM. If HE does not call us and give us the will, we will not choose HIM because we prefer darkness rather than light. Being in GOD shows up all our flaws. This feeling is very discomforting and some prefer not to know the extent of their flaws. We must therefore be willing, even with our flaws, to submit to HIM.

'Light' shows us who we really are and, being so corrupt, we do not want to see ourselves for what we really are relative to 'ABSOLUTE TRUTH'. You will come to this conclusion when you begin to understand one of the attributes of GOD, which is TRUTH. You will only know this after HE reveals himself to you. HE is nothing like you could ever imagine. HE is so much more that it overwhelms you. A typical response is to break down sobbing.

We have already been provided all the information we need to come to the conclusion that we all need GOD'S assistance and without HIS help we are lost and will remain lost. The way to HIM is outlined in HIS Holy Word, the HOLY BIBLE. What we need to do is, repent of our sins, submit to HIM and HE will guide us back to the path of 'TRUTH'.

I mentioned above that we have free choice. That is the truth. However, if GOD needs you to go in a certain direction you cannot resist. It may even appear to you that it is your choice but you would be mistaken. HE is the one that gives you your unique qualities and HE has absolute authority over you and is able to mould you as HE wishes. HE is GOD and with HIM all things are possible. Whatever needs to be done to execute HIS plan, will be done.

HE has the power to harden or soften your heart so you should pray to HIM to soften your heart so that you will hear HIM and listen to HIM when HE speaks to you.

When HE first called me, I was very unsure of how to react because I did not understand what was happening to me. It was not rational. It was supernatural and so defied logic as we know it. However, deep down I knew it was the right thing to do. I felt at peace with what I was about to do. I felt an inner peace and joy that was not of me. It was of someone in whom I could trust and have complete faith. I was now in the presence of a powerful, supernatural force and knew at that instant that following HIM would be the most important decision of my life. Ever since that moment, I have never felt alone and always have that inner peace.

Having had my conversion experience, I thought it would fade with time. However, the opposite is true. It was just the beginning of a continual process of getting to know GOD and understanding HIS character. It is a transformational process. Your conversion process may not be the same as mine as GOD is diverse in HIS ways and your conversion experience is designed for you only. It is an intimate and very private experience.

To help me in this process of transformation, I have continued attending the church to which he sent me and I still do. If it were left to me alone, I know I would not have gone to that church and continued to do so, since then.

Sometimes, when driving to church, I would ask myself, why am I doing this? But I knew the answer. I really wanted to know more about the GOD that read my mind and answered my prayer and still continues to guide me.

I am always aware of HIS presence and ask HIM questions about life and my purpose here. HE guides me daily and although

I may have many trials in this life, HE has promised to be with me through them all. HE has never disappointed me.

> Bible quote - (Matthew 28:20- 'Behold I will be with
> you always, even to the end of the age')

One important thing that I learned is, if you have not been converted, you do not know HIM and you do not hear HIM when HE speaks, because the HOLY SPIRIT is not in you. You have to repent of your sins and ask HIM to reveal himself to you and guide you. Only then will you begin to know and understand the one true GOD. Don't try to imagine who GOD is; let HIM show you HIS true self.

> Bible reference - (John 10:27,28 'My children know
> my voice and they follow me')

GOD does not change. We discussed this previously but I will now explain further. Everything in creation changes relative to everything else, but HE does not change and will never change. HE is the absolute reference, HE is perfect, HE does not need to change. This means that you can depend on HIM to keep his promises. HE makes a covenant with you, at the time of your conversation, at which time HE starts the process of transforming you into a new person, the one HE meant you to be. When you see HIM operate in your life and you look at HIS documented words on how HE communicated in THE BIBLE, you will discover that HE is the same GOD of THE BIBLE.

'HE is GOD and there is no other'. HE makes all the decisions regarding heaven and the earth. That does not mean that if you ask HIM for something specific HE will not give it to you, if it is for your good. This is also part of HIS plan. Being your FATHER, HE knows how to give good gifts to HIS children. This is one of the reasons why HE encourages you to pray to HIM. HE wants you to communicate with HIM. Once you are adopted by HIM, HE becomes your SPIRITUAL FATHER.

Bible quote - Good gifts (Matthew 7:11)

Do you understand the honor it bestows on you to be a child of GOD? The privileges it affords you? This is the ultimate experience for any human being and no other gift supersedes it. It is the most important thing that could happen in your life. I know this because I have experienced it and continue to benefit from the ongoing experience.

We know that GOD THE FATHER is the ultimate decision maker regarding all things of HEAVEN and EARTH. We see it in everything around us because it is of one mind. In addition, HIS SON JESUS, told us so. JESUS says that HE only does the FATHER'S will and HE and the FATHER are ONE. HE knows the Father's will. The HOLY SPIRIT also does the FATHER'S will.

The fingerprint of GOD is an integral part of the universe. HIS intelligence, HIS character, HIS singularity. There can only be a single, intelligent reference at the source of the universe otherwise we would be in total disharmony. Life would not exist as it does, with the human body in harmony with itself and the environment. The universe also exhibits this harmonious trait. It is all coordinated and unified indicating one controlling force. Any disruption of this harmony is a result of SIN.

Our body has several subsystems, all operating in total harmony, which also indicates that a single, intelligent source designed it. Two or more separate, independent entities cannot be in harmony with each other unless they are designed to do so. Two separate controlling forces would have conflicting goals unless fully coordinated. The universe and everything in it act as one which confirms that one unifying force or intelligence designed and controls it all. The fingerprint of GOD is the fabric of the universe.

If there were two or more independent GODS, there would be two or more controlling forces. This could not work. There can be only one final decision maker in the design, creation, and functioning of the universe and everything in it. This ultimate controlling force coordinates and choreographs all events and has the final say in everything. In reality and for all practical purposes, there needs to be a single controller just like HE tells us. GOD THE FATHER, GOD THE SON, and GOD THE HOLY SPIRIT have separate responsibilities and there is no conflict.

All the knowledge and all the intelligence that will ever be already exists in GOD'S world. It is all that there is and all that there needs to be. This is the prefect world that JESUS calls PARADISE. It is perfection defined. It gave rise to our universe but our universe is now flawed as a result of our doing, we SINNED! We disobeyed GOD'S spiritual laws the consequence of which is death. This is total separation from HIM, the true reference. But there is Hope! HE has given us a way to redeem ourselves!

The Trinity

ENTITIES. There are 'GOD the FATHER', 'GOD the SON' and 'GOD the HOLY SPIRIT'. If this had not been revealed to us we could not have deduced it by logical reasoning. This is a supernatural concept and so not within our reasoning ability.

One very important attribute of this relationship is that, even though THEY are separate ENTITIES, THEY work together as ONE. THEY each have separate responsibilities all aimed toward one goal, creating and maintaining a PERFECT EXISTENCE.

THEY existed before time, in fact, THEY created time. We are not able to perceive such an existence as our perception is limited in time. THEY exist outside of time.

Physicists have determined that time is relevant only to our universe. This is consistent with what GOD has told us about HIMSELF, being ETERNAL.

THE TRINITY is a perfect union. There is no conflict. The THREE MEMBERS execute THEIR responsibilities flawlessly.

GOD THE FATHER is the decision maker whereas THE SON and THE HOLY SPIRIT execute the FAFHER'S will. Once you are converted, the HOLY SPIRIT is sent to guide you in the ways of the FATHER so that you are transformed into the image of his SON. This transformation process may be slow, so don't get discouraged that sometimes it is as if no change is occurring. Once you have been converted you can be assured that the transformation process-will continue unto completion. You become one of the chosen few and will remain HIS eternally.

Knowing God

What I am trying to convey to you is not about religion but about developing a direct relationship with your 'CREATOR'. Even if you do not believe such a BEING exists and would have the desire to communicate with you, as in my case, 'Just do it'. Ask HIM to reveal himself to you and guide you throughout your life. You have so much to gain from asking. Is it not reasonable to deduce that, if there is a GOD, this ENTITY must be able to communicate with you? Having reached that conclusion the next step should be an attempt to communicate with HIM.

Since you do not know HIM or where to locate HIM, you must ask HIM to reveal HIMSELF to you. The way is described in HIS holy book, THE HOLY BIBLE.

Trust me, you don't have to say it out loud. I did not! Yet HE read my mind and responded.

GOD speaks the universal language that we all understand but of which we are unaware. HE can contact you any time HE wishes. HE has given us his 'WORD' (THE HOLY BIBLE) and wants us to willingly come to HIM.

GOD initiates the relationship and this is part of HIS plan. The way is described in HIS book, THE HOLY BIBLE. If HE did not influence us to turn to HIM we would go in the opposite direction and continue in 'darkness'. However, once you submit and HE chooses you, you cannot change your mind. When you fully understand what is happening to you, you do not want to resist, you willingly submit. You then begin to wonder how you lived your life before, without HIM. You then humble yourself and submit to HIS guidance. After the initiation of a relationship with GOD, you will come to find it to be the most satisfying relationship you will ever have in your lifetime.

When you are called to be in a relationship with GOD, you can only enter HIS presence if you are pure. Since none of us is pure, a sacrifice of absolute purity had to be made. This is one of the requirements of a 'JUST GOD'. JESUS, became the LAMB, the perfect sacrifice to save us. JESUS, one of the TRINITY, had the purity to save us.

While you are still on this earth, the HOLY SPIRIT, the COMFORTER, is with us to help lead us to all TRUTH. You will, however, have the opportunity to be with GOD when you pass from this world and are fully JUSTIFIED.

When you read GOD'S word, you begin to develop an understanding of how HE communicates with us. Some of the ways HE communicated with people thousands of years ago are some of the same ways HE communicates with us today. HE communicates with us directly, as well as indirectly, through other people. In my case, HE told different people the same message or different parts of the same message so that when it got to me, it was communicated completely, and with redundancy. This ensured there were no errors in communication. When all put together, you understand the full message. In each case the Person interceding is the HOLY SPIRIT.

It is not possible for nonbelievers to understand this process because it is a supernatural process. I know this because I was once a nonbeliever. When I first read the Bible, it was just a story to which I did not relate. After my conversion, its whole meaning changed. Now, when I read the BIBLE, my FATHER is speaking to me through his word.

One of the most comforting things of being one of GOD'S chosen children is the 'peace' that HE gives you even in the most difficult circumstances. It is truly a 'peace' that passes all understanding'. It is a supernatural 'PEACE'.

GOD always keeps HIS promises. You can therefore depend on HIM to be with you through all your trials.

GOD made man in HIS own image. This includes our emotions and our ability to create. Remember, HE created an ordered system and this is an indication of HIS character.

To help us understand who GOD is, we can examine ourselves, our approach to creating things and our emotions. The difference is that HE is perfect and all of his character traits are designed to maintain order.

GOD tells us, and it is also my belief, that everything comes from and is of HIM. Without HIM we are nothing. We would not exist. Something happened to break (we sinned) our direct connection to HIM and as a result, we are lost.

> Bible quote (John 1:3, KJV - 'All things were made by HIM; and without HIM, there was nothing made, that was made')

You cannot find GOD because you don't know who HE is or where to find HIM. Can one even begin to look for someone if you do not know who you are looking for or even where to start? GOD has to reveal HIMSELF to you. Otherwise, HE is completely out of reach. Remember, alone, without GOD, we are lost and without the 'TRUE' reference. If you are yourself lost, how can you find someone whom you do not know?

I do not believe that, if you ask, HE will not respond to your request to reveal himself to you. HE may even reveal HIMSELF to you before you ask if it is a significant part of HIS plan! When you read about Paul's conversion, in the bible, he was not given a choice. He was persecuting GOD'S people and was struck down, made blind and converted in an instant. GOD had a specific purpose for him in HIS plan, to become one of HIS servants and be one of HIS advocates. Paul was uniquely qualified and so was selected without question.

GOD may also answer your prayer when you ask or some time later. HIS response is based on HIS plan for you. GOD knows your heart and this is how you are judged. If you genuinely repent of your sins and ask GOD to forgive you and guide you, it is because HE has already selected you for conversion and the process has been initiated.

GOD is merciful and therefore willing to save us. We also can be merciful. But, because of our SINS our mercy is flawed. HE is a JUST and perfect GOD, HE knows our hearts and so will provide the perfect response when we pray to HIM. We are far from perfect and so should not even try to predict what HIS response will be. What I do know is that HIS response will be 'JUST'. If you are sorry for your SINS and want to make a genuine effort to do good (repent) HE will respond favorably to your request.

> Bible quote (1 John 1:9 - 'If we confess our sins, HE is faithful and just to forgive us our sins and cleanse us from all unrighteousness')

Just like JESUS spoke to us in parables or symbolic terms, when the HOLY SPIRIT speaks to us it can also be in symbolic terms. Like the vine and the branches, if we are not connected to JESUS we will perish. Everyone can understand that. When GOD said to man "Be fruitful and multiply", fruitful means bear a lot of fruit or have many children.

ABSTRACT REPRESENTATION OF GOD

> Bible quote - (John 15 - 'I am the vine and you are the branches')

I have included an abstract representation of creation which was custom painted by a true believer, Jack Sependa, at my request. I will also interpret the painting since it may not be immediately apparent to most readers. (See painting on previous page)

GOD calls HIMSELF "THE ALPHA AND THE OMEGA". GOD is the center of the universe. HE is surrounded by white light. God spoke creation into existence as shown by the exploding colors representing creation. Everything is of GOD. Light connects the universe and it is no wonder JESUS calls himself 'THE LIGHT OF THE WORLD'.

This painting was done in one day. Jack has a unique approach to getting inspiration before he starts each painting. He puts a symbol of the cross on the blank canvas and then prays to GOD for inspiration. The painting was inspired by the HOLY SPIRIT.

The HOLY SPIRIT also revealed a prophesy to Jack, which he relayed to me in real time as we sat for breakfast at a Hotel in Santa Fe. I knew it was real because, as we were talking, he suddenly stopped and tears came to his eyes and his words were, "John, the HOLY SPIRIT is giving me a message for you". It is this message that is outlined in this book. I will tell you the full story some time in the near future.

God's Fingerprint

Everything we do, say or think is ordered if it is in harmony with the universal order. This principle applies to all aspects of the universe. If we extrapolate, we can conclude that the universe, being ordered, originated from the same intelligent source. We all, therefore, originated from the same fixed reference from which everything originates.

'Absolute Intelligence' and 'Never Changing' are two of the attributes of GOD. It is therefore much more logical to conclude that

GOD created the universe rather than a 'random' occurrence such as the evolutionist's 'Big Bang'. 'Random' is foreign to the order which we see and to life as we know it.

GOD stays in contact and in complete control of the universe and everything in it even if you may not believe this to be true. HE intervenes when necessary otherwise we would be completely out of control. None of us has the knowledge or ability to maintain order in such a complex system because there can only be a single source of control. We cannot each be a reference as this would lead to confusion. There can only be one ABSOLUTE SPIRITUAL and MATERIAL reference. One fixed reference. One perfect reference. Looking at these requirements we can see that none of us qualifies as we are all imperfect.

GOD'S attributes are reflected in the universe. He is our rock, our reference, HE tells us so and HE does 'NOT CHANGE'. This is critical for creation and in any attempt to construct or fabricate an ordered system using the materials available to us, matter, as we know it.

Based on what we have already discussed, is it not reasonable to conclude that this universe and everything in it must be of intelligent design?

Only a source with intelligence can create an ordered system. Intelligence defines order. This is the only logical conclusion.

Spiritual Laws

'MAN'S' current spiritual nature is such that we do not have the ability to live peacefully with each other. Disharmony begins with family conflicts and spreads to communities, cities, countries and the entire world. 'MAN' cannot break 'MATERIAL' laws, but can and does break 'SPIRITUAL' laws.

To try to maintain order in our world, we develop and implement laws relative to our conduct and a means to enforce them. These laws are based on standard ethical practices that all are expected to obey. These are not 'MATERIAL' but 'SPIRITUAL' laws based on BIBLICAL principles. They are behavior based. Because of SIN, if we did not have laws, society would end up in a state of anarchy.

There will always be some of us who are unable to comply with the laws of the land and have to be disciplined. This is an indication that we are flawed.

In order to maintain absolute order, no disorder can be allowed. The laws of GOD are much more strict than the laws we have developed. None of us, on our own, is capable of obeying all of GOD'S laws. HE created a perfectly ordered system but, through what HE calls SIN, we introduced disorder, and became separated from HIM and HIS guidance.

The first created human being, ADAM, made the choice of eating from 'The Tree of Knowledge of Good and Evil' so he became aware of good and evil. In so doing, he introduced disorder as he now had the ability to choose evil, which he did and we all continue to SIN.

God made us in HIS image so we can make our own choices except the one which HE withheld from us. HE gave us direction as to the path we should take, we disobeyed and are now suffering the consequences. We lost our absolute reference and as a result, lost our way.

We have discussed how a fixed reference is critical to an ordered system. Once we were disconnected, we substituted our own references which is not universal and absolute. We have not only ourselves made our own references, but we sometimes keep changing the reference. We can only have harmony among ourselves if we are following the 'ABSOLUTE SPIRITUAL REFERENCE' which is fixed and never changes. When we each have our own reference, it is impossible for us to live together in harmony (GODLY). We become a disordered system. We introduce negativity into our behavior, hence disorder. We are still intelligent beings so we can create order, but we now choose to create disorder in our society, resulting in disharmony.

Our FATHER does not tolerate disharmony as this is not a part of his nature. HE is a 'MERCIFUL' GOD and so HE has given us a way to redeem ourselves.

Since HE is also a GOD of 'JUSTICE', it was not possible for HIM to forgive us without a sacrifice. Remember, the consequence of SIN is death. This is one of GOD'S SPIRITUAL LAWS. Since we now have a degree of disorder within us, none of us is capable of making this correction on our own. Once a degree of disorder comes onto a SPIRITUAL system, the only way it can be corrected is by making a 'SPIRITUAL SACRIFICE'. This satisfies 'SPIRITUAL JUSTICE' which mustbe upheld. HE therefore sacrificed a part of HIMSELF, HIS son JESUS, as none of us was pure enough to be or to provide this sacrifice. HE did this because 'HE LOVES' us.

Satan

SATAN is a fallen ARC ANGEL who was cast out of heaven, to earth, because of his rebellion against GOD. He is a negative (Evil) force that works to create disorder in our spiritual realm. He is against GOD and his people. He attacks us by trying to influence the way we think and reason, and so tries to manipulates us into doing his will. He tempted and influenced MAN to SIN against GOD and once this happened he then had power over MAN.

SIN is the result of this negative force in our lives. It destroys anything good and ordered and so must be stopped. Its aim is to destroy anything GODLY because GOD represents all goodness and order. If we are believers in JESUS, SATAN only has power over us when we give it to him through SIN. By ourselves, we do not have the power to overcome such an evil force so we have to rely on GOD to help us.

When JESUS died for us on the cross, JESUS became our SIN (see Bible quote below) and believers have been given power over SIN and SATAN. Jesus destroyed the works of SATAN and took his power away from him. HIS people now have power over SIN, but can still sin because they are not yet perfect and still possess part of their old selves. They are still tempted and sometimes make the wrong choice. They will not be perfect until they are again fully joined with GOD.

> Bible quote - (2 Corinthians 5:21 - 'For HE made HIM who knew no SIN to be sin for us, that we might become the righteousness of GOD in HIM')

The Bible tells us that the thief (SATAN) came to steal, kill and destroy but that JESUS came that we would have life and have it more abundantly.

Man

The TRINITY has always existed. The FATHER designed and planned creation together with his son JESUS and the HOLY SPIRIT. They have always and will always work in harmony. MAN was the subject of creation. Everything else was created to accommodate us.

MAN was made in GOD'S image. This means that MAN was given similar attributes to GOD'S. Some of the attributes given to man are creativity, love, empathy, mercy and even hate. The difference is that GOD uses HIS attributes in accordance with HIS holy laws in order to maintain harmony in heaven and on earth. HIS attributes are absolute in all respects.

MAN was given free choice but was also given a limit as to what he was allowed to do.

However, man sinned and was disconnected from GOD. In so doing, we lost our moral reference and became morally and spiritually corrupt. SIN is rebellion against GOD'S SPIRITUAL laws, thus creating moral and SPIRITUAL disorder. God cannot tolerate SIN because it disrupts SPIRITUAL order. Those that SIN are rejected from HIS ordered world. SIN is punishable by death. But, even in our SIN, GOD still loved us and sacrificed HIS SON to save us from eternal death. Eternal death is eternal separation from HIM. JESUS willingly gave HIMSELF as the sacrifice to save us from eternal death.

GOD is also a GOD of 'TRUTH and JUSTICE' and made laws designed to uphold spiritual order that must be obeyed. If we disobey them, we will most certainly suffer the consequences. This is where we come face to face with the fact that GOD has absolute authority over all things and commands our awe, love and respect.

Order cannot survive with disorder active in its midst. The only way to eradicate disorder is to attack it with a perfectly ordered system. JESUS was that perfectly ordered system.

In any ordered system, there can only be one ultimate leader. GOD is that ultimate leader.

When man first sinned, he immediately inherited the death sentence. All of us being direct descendants must also suffer the same fate, as the effect is systemic.

Fortunately, GOD is merciful and has given us the opportunity to be accepted back into HIS fold, if we confess our sins and commit to HIS guidance in our lives. We must have faith that he loves and wants the best for us.

With all this in mind, we should realize that HE did not have to create us. He did not have to give us the opportunity to live, know that we 'ARE', be able to interact with our environment and to give and receive love.

HE was perfect without us, but wanted to share HIS 'WONDER' with us. This is the greatest gift you could ever imagine.

Some say that GOD'S appreciation for loving and sharing comes from the fact that HE has always shared such a relationship with HIS SON and the HOLY SPIRIT. THEY are three separate entities but act as one. EACH ONE has a separate responsibility but act as an integrated unit.

We are eternal beings but in order to spend eternity with GOD, we have to repent of our SINS and commit to following his guidance. Otherwise, we will be eternally separated from HIM. I don't think it takes much debate as to which option we should choose.

GOD has shared intelligence and knowledge with us. But GOD has the power to conceal knowledge and wisdom from us, if HE so chooses, and HE sometimes does. No matter how intelligent you think you are, you can never discover certain truths unless HE chooses to reveal them to you. HE will conceal such knowledge from you until HE is ready to reveal it to you. This is because HE is and will remain in control. HE has a plan and all revelation will be done in accordance with HIS plan.

Each of us has a purpose in life. The ultimate goal is for goodness and 'TRUTH' to prevail. However, bad things do happen but in the end, everything, even the bad, now happen for the ultimate good.

We are each designed to perform different tasks over a lifetime and are given special gifts to perform these tasks effectively and efficiently. GOD has a plan for each of us and has given us the necessary gifts to carry out HIS plan. HE coordinates our actions to fulfill HIS overall plan for the world. HE remains in total control. We cannot change this plan as we do not have the power to alter it.

Since we are lost, we create our own order which may not be in harmony with the universal order. When we chose our own path, by detaching ourselves from our source, we became lost. Once lost, we have no true reference and it is impossible to find our way back. Only GOD, who knows the path, can direct us back.

If we examine what it is like to be lost, unless someone who knows the way directs us back to the true path, we cannot find our way. To find the true path, we either have to be lead back by someone who knows the way or find a reference, on that path, which we can use to guide us back. GOD is the only one who can lead us back to the true SPIRITUAL REFERENCE. It is impossible to find the reference as we do not know what to look for on whichever path we may travel. Even if we saw the true reference we would not recognize it. GOD must reveal it to us.

God made us in his own image and gave us free choice but limited our knowledge and intelligence. How could he have given us all his power knowing we could not be trusted once we were given free choice and access to the knowledge of good and evil.

Since we had nothing to do with our being here, we have no reason to be proud of our intelligence and ability to accumulate knowledge. Whatever we are able to accomplish are gifts freely given to us.

Jesus

Jesus is the one and only SON of GOD. When man sinned, the only way he could be saved was if a 'SPIRITUAL' sacrifice was made to satisfy GOD'S requirement for JUSTICE. There was nothing in creation that could meet these requirements because we are all tainted with the curse of 'SIN'. Only the prefect sacrifice was good enough and JESUS willingly became that sacrifice. He still loved us even though we were now SINNERS and no longer loved HIM.

GOD never stopped loving us even when we did not love HIM. HE was willing to sacrifice HIS SON to redeem us and JESUS was willing to be the SACRIFICIAL LAMB. This is the true definition of unconditional love.

When I read the Bible, I always consider JESUS'S quotes to be profound and with much more depth than initially meets the eye. We must always remember that HE is part of the TRINITY and so sometimes speaks in absolute terms. HE speaks universal TRUTH. HE was there in the beginning and so participated in everything that was made. When HE makes a statement such as 'I am the (spiritual) VINE and you are the branches, without ME you cannot live'; This is absolute TRUTH, in every sense. The branches of a tree must always be attached to the vine. The vine or trunk supples the nutrients of life. The same applies to the human body. The body parts must all be attached otherwise they die. Earth supplies water through rain, rivers and streams; otherwise, life could not exist on our planet. The elements of life must be constantly flowing in the body in order to support life, any disconnected part will die.

Jesus is our only hope for eternal life as HE died for us so we can now be reconnected with GOD through HIM. The closeness of the relationship between the GODHEAD and us is seen in the way

JESUS refers to us once we are saved. JESUS tells us HE is now our 'BROTHER' and GOD is our 'FATHER'. This is an indication of who we are in the family relationship with the 'FATHER' and the 'SON', once we are adopted (SAVED) by the FATHER. But, to gain this honor, we have to be BORN AGAIN not of flesh but of the 'SPIRIT'.

GOD tells us that the first step to redemption is to accept HIS SON JESUS CHRIST as our savior, repent of our sins and commit to following HIM. GOD then reveals HIMSELF to us directly (through the HOLY SPIRIT) and through HIS 'HOLY' book, 'THE BIBLE'. However, in order to properly interpret the BIBLE we need the HOLY SPIRIT to guide us in understanding it. It is impossible to understand the full meaning of the BIBLE without the help of the HOLY SPIRIT. But GOD'S word, as manifest in the BIBLE, is how we are initially made aware of HIM and are converted.

After you are converted, you become one of GOD'S chosen people. HE cares for you and nourishes you like a mother would a child. HE sends you the HOLY SPIRIT, the HELPER, COMFORTER, SPIRIT OF TRUTH to assist you in knowing and understanding HIM. This is what is referred to as being 'BORN AGAIN'. You are now born of the 'SPIRIT'.

Once this happens in your life, this means that you have been selected to be a SPIRITUAL CHILD of GOD and GOD THE FATHER is now your 'FATHER'. From this point onward, you are forever saved and no one can take you from JESUS'S hands. GOD selects you and gives HIS SON, JESUS, authority over you. JESUS is now given the responsibility to protect and care for you and you are guided by the HOLY SPIRIT. Your life is changed forever.

Bible quote - 'No one will snatch them out of my hand' (John 10: 28-30)

A transformational process now begins in your life which will continue until you are fully made into the image of JESUS. While you are still here, the process will not be complete, but will be when the FATHER calls you to be with him and you are fully JUSTIFIED.

If we consider this objectively, we see that GOD sacrificed HIMSELF to save us since GOD and JESUS are one. JESUS humiliated HIMSELF, being cursed by us and put to death on a cross. During this period 'JESUS' was separated from the 'FATHER', giving up HIS power as a GOD for 'SINFUL' man.

JESUS lived with us from birth to mature adulthood, experiencing our world first hand and overcoming all its temptations. HE is the only one that has been able to do this in human form. HE was tempted much more than we are but as HE so aptly said 'I overcame the world'. HIS promise to us of eternal life depended on the fact that HE overcame the world.

Bible quote (Jesus overcame the world- John 16:33)

Overcoming the world also means that HE overcame DEATH. This is important to understand because of HIS promise of eternal life. The only way HE can make such a promise to us is if HE has power over death, which is the only thing that separates us from eternal life. Having died for our SINS, JESUS was given complete authority over us. Even though HE is our brother, HE is also our KING and commands our respect.

Jesus rose from the dead on the third day after being crucified, just as HE prophesied. HE now sits on the right hand of the 'FATHER' until HE comes back for HIS chosen people.

Sin

SIN creates disorder. It is systemic. It not only affects our thought process but also matter, the material world. SIN results in our separation from the ABSOLUTE (spiritual) reference. GOD is the absolute reference and had no choice but to separate us from HIMSELF as a result of our breaking HIS spiritual law. By SINNING, we caused a disruption in the spiritual realm to which we were directly connected through HIM. This is why we had to be separated from HIM, as SIN has no place in that perfect world.

The order given to man was clear. 'Do not eat from 'The Tree of Knowledge of Good and Evil.' But man did eat from that tree and as a result was cursed, lost the absolute reference (GOD) and substituted it with his own limited reference which is in disharmony with universal order.

God is completely and totally ordered and made us in HIS image. However, we chose to introduce disorder (SIN) by disobeying HIS command. Once this disorder was initiated, it spread like a cancer affecting everything and everyone with which it came in contact. It is a destructive force. It is non-productive and has no place in an ordered world. SIN'S only function is to create disorder and so it is unacceptable in an ordered world.

As indicated above, SIN has affected not only our minds, in the way we think, but also matter as it exists (in ordered systems) in our universe. You see, GOD also cursed the earth, and SIN is systemic. What was initially ordered and perfect is now tainted with disorder.

HE is a 'JUST GOD' so HE had to sacrifice part of HIMSELF to redeem us and he made a covenant with us to again accept us as HIS own. Since we could not redeem ourselves, the sacrifice was HIS SON JESUS.

Love

There are four types of love described in the Greek language.

These are: Storge - Empathetic love

Phila - The love of a friend

Eros - Romantic love

Agape - Unconditional love

The love I will address is 'Agape' or Unconditional Love, which is the love that GOD has for us. This LOVE is the most powerful and binding force in the spiritual world. It is a spiritual law. It is a commitment to maintain a covenant relationship with another which does not change (it is unconditional). Love, the emotion, is secondary as this is subject to change based on the emotional state of the parties involved.

UNCONDITIONAL LOVE incorporates the SPIRITUAL reference that binds us all together in perfect harmony.

Treat another individual as you would have yourself be treated. We should love one another in spite of challenges we may encounter in the relationship and not only because of the benefits we will get from it. This love is self sacrificing and will make sacrifices to ensure that order is maintained in the relationship.

(1 Corinthians Chapter 13 defines unconditional love)

LOVE is an order producing force. This is the reason why our GOD has told us that 'Of the three, FAITH, HOPE and LOVE, the greatest of these is LOVE'.

Bible quote - (Greatest of these is love, 1 Corinthians 13:13)

Our secular definition of love is selfish. It is self serving and so not the same as the 'LOVE' GOD has for us. When we say we love someone, we typically mean that the individual satisfies a selfish need in us. Without them we are unfulfilled. As we examine this concept, we see that our definition of love is not giving but receiving or taking. In other words, expecting something in return for the 'love' we give.

GOD'S LOVE gives not requiring anything in return. Similar to the love a mother feels for her child. It is instinctual and ordered. For order to be maintained, we must overcome outside influences and use only the 'TRUE', absolute reference as a guide. LOVE is an ordered system. One should love not because of the benefit to us but because of the benefit to others. The benefit to you will follow. It is the natural order of things. Remember, GOD still loved us even after we SINNED.

Love is a conscious decision that we make that sometimes sacrifices our own happiness. This is necessary to ensure that order is maintained in the universal system. This is critical in our fallen world. Each part or member of the universal system must sometimes make sacrifices for the whole.

In our fallen world, this is essential for order to be upheld. It therefore seems logical that, to save us and help us return to order, GOD had to sacrifice HIMSELF in the form of HIS SON. This is what 'UNCONDITIONAL LOVE' does. It is self sacrificing.

No human being could ever have come up with such a definition of love as we typically think selfishly. If we look at the 'Big Picture' of us being the ultimate ordered system in our world, it becomes clear that this LOVE is critical to maintaining order. It binds us together, making us unified. Its cohesive force is necessary for our survival. If there is no love, there is no unity. If there is no unity, we become self destructive and will become even more disordered and lost. We will have become totally detached from our universal reference, on our own, and unable to get back on course.

In order to stay on course, we need to understand the definition of UNCONDITIONAL LOVE and practice it. This is not something we can do on our own because most of us are already lost and look at the world from a selfish perspective. We may think this selfish perspective is natural but it is not conducive to the universal order. We need the guidance of the absolute reference, but first we must recognize that we are lost and need help.

The only ones that will be redeemed are those who recognize the need for help and ask for and receive guidance. Even after conversion, we still need constant guidance and support to keep on track. It is a daily battle and we must depend on continual guidance from our ABSOLUTE REFERENCE, our HEAVENLY FATHER.

When GOD created us HE created a need in us that only HE can fulfill. Things of the world will never satisfy this need.

Trust

Trust is a concept that is so delicate it takes only one mistake or act of mistrust for it to be completely lost. Trust develops from one being consistently trustworthy and so it is earned. Someone we trust is someone we can depend on to look objectively at the situation at hand and judge impartially. Such a person can be depended upon to tell the truth and be fair in every situation. Once an individual is given the honor of being trustworthy, he or she is used as a standard and expected to continue acting in character. Any change in this individual that questions this attribute can result in trust being lost. (Always remember, GOD does not change and so he can always be trusted)

Trust can sometimes take years to build but can be lost with a single negative act. Consistency is critical for trust to be maintained. This is another example of how important it is for us to live by unchanging, upright standards and be a 'light' to all.

If our reference is not absolute, there is always the chance for error. The only absolute reference is GOD. GOD never changes. We can absolutely trust HIM. For us to trust someone, it is critical that the person remains the way we perceive them. We develop trust over years of association with someone by analyzing their actions and responses in various situations. Trust is very fragile and so we should always be aware of how we are perceived and uphold the highest ethical standards.

God's Forgiveness

GOD promises to forgive us if we repent of our sins, humble ourselves before HIM and commit to keeping HIS commandments. This is necessary in order to recreate the order that was disrupted. There is a consequence for breaking spiritual laws and a sacrifice necessary to redeem us. Forgiveness is only possible if a suitable sacrifice is made. This is GOD'S JUSTICE and is one of HIS attributes.

There had to be a sacrifice for HIM to again accept us as one of HIS own. None of us is worthy to be that sacrifice as we are all impure. HE therefore sacrificed a part of HIMSELF for us by sending HIS SON to die for our sins (GOD and JESUS are one). There is no sacrifice we could have given HIM that would have been sufficient because we are all unworthy. Only the perfect sacrifice would be acceptable to a JUST GOD. We do not qualify and nothing we posses qualifies.

Only a GOD of LOVE, MERCY and JUSTICE would consider doing this for us. HIS SON, also part of the Godhead gave HIMSELF to be that sacrifice. This is a very powerful and humbling acknowledgement.

Our Forgiveness

When we forgive someone of an offense against us, we are not only forgiving the person but are also freeing ourselves from dislike, hate or contempt for that person. Holding resentment against someone that has offended us hurts us more than it hurts the person who committed the offense.

This is one of the reasons why forgiveness is so important as it provides us relief from the burden of dislike, hate, and contempt.

Prayer

Prayer is direct communication with GOD. We know that GOD has a plan for us and a plan for creation but HE still encourages us to pray to HIM. In prayer, we can ask for blessings even regarding specific things. HE will grant us such requests if they are things that are 'GOOD' for us.

Even though GOD has a plan for us, HE will listen and respond to our prayer requests. HE gives gifts to HIS children and, being our father, HE knows better than we do how to give good gifts to HIS children.

Bible quote (Good gifts - Matthew 7:11)

Our prayers should always include requests for wisdom so that we will better understand what he desires of us. But most of all, we should recognize and address GOD'S sovereign power over us. Even JESUS said, 'Not MY will but YOUR will be done'

Bible quote 'not my will (Luke 22:42)

The more we pray to OUR FATHER, the faster and deeper our relationship with HIM will grow. When we pray we should believe or have faith that our prayers will be answered. GOD will always answer our prayers but the answer may sometimes not be what we expect. Remember, HE only gives us things that are good for us or for the universal good. HIS response may be delayed so you must be patient. HE does everything in HIS own time in accordance with HIS plan. Sometimes HE really blesses us and gives us even more than we expect.

Always remember that prayer is direct communication with GOD and the best way of getting to better know and understand HIM. Also, remember that you are in communication with 'GOD' and must remain humble. You are HIS child and HE loves you. In this context, you can talk to HIM about anything and should. HE will always look out for you and give you good advice.

Faith

Bible reference – Hebrews 11:1' Faith is the substance of things hoped for, the evidence of things not seen.'

Faith is believing in something you cannot see or even rationally explain. However, you believe because of GOD'S supernatural influence.

True faith is not initiated by us but by GOD. We know that it is supernatural because it does not fade but gets stronger with time as our relationship with GOD grows. It is not something that can be rationally explained. Yet we know it is real and we can put trust and have confidence in its source.

FAITH cannot be fully understood by someone who is not a believer. Typically, non believers argue that unless you can see it and touch it, it is not there, it does not exist, it is all in the mind.

Having true faith is an indication that one is in GOD and the HOLY SPIRIT is in you. A measure of FAITH is one of the gifts given by GOD when you become a believer and you are adopted as one of HIS chosen children.

With this FAITH, you have no doubt that GOD exists and that HE created the universe and everything in it. When you read the Bible, you believe everything HE says about himself, what HE says about the world and what HE says about you. You will find that HE tells you things about yourself that you find later to be true. One can only conclude that, if HE knows more about you than you know about yourself, then HE must have created you. Your faith then grows even deeper when these truths are revealed to you.

Hope

Hope is the desire that things wished for will come true. If we pray for something, we hope it will come true. But, it is up to GOD as to whether or not HE will grant you your desire. You should always remember that HE will only grant you things that are good

174

for you or contribute to the universal good. We may not always ask for things that are good for us so don't be disappointed if you do not get the response you desire.

Even when bad things happen, there is good that comes out of that experience or event. We should remember that we are not in a position to see the big picture, but HE is. Things will always go according to HIS divine plan and we should not question it.

Parents, understand that your children are 'from' you but not 'of' you. You do not own them. They are all GOD'S children, as are you. Their destiny is up to HIM and you need to respect HIS sovereign will.

The fate of your child is not for you to decide. It is for GOD only to decide. I know this is difficult to accept but this is the truth. You may believe you know what is best for your children and you do not want to see them harmed or suffer but, ultimately, it is GOD'S will that will prevail. Accept that His will provides the best outcome even if you may not understand it now.

With the above in mind, we can still hope that what 'we' consider all good things will be given to us. We should always remember that we live in an imperfect world and will experience suffering. But, if you are one of GOD'S chosen children, HE will always be with you through any difficulties or trials this life presents.

Final Message

There is a fixed reference that connects all of us and everything in the universe. This fixed reference is an intelligent source. The intelligent source is God.

The only reason I have been able to write this book is because I have a fixed reference, the true reference, my FATHER GOD. My focus was always on HIM to guide me in every thought and word. Whenever I had a problem with expressing a thought, I would ask HIM to tell me what to write to convey the intended message. In fact, HE has guided my thoughts throughout this process, from beginning to end. My hope is that this book has reflected HIS divine guidance and offers some more clarity as to your purpose in this world. Keep in mind that I am human, but I think I have accurately recorded the basic principles I am trying to convey to you, the reader.

Throughout my life I have been in training to write this book. Looking back, I now see the complete picture as all aspects of my life are relevant to the included message.

The idea for this book was given to me by my 'FATHER GOD'. HE sent me the message that I would have to do nothing and 'IT' would all come to me. That was so true as all I have written have been my life experiences and training as an engineer.

HE showed me how to look at the 'big picture' and put it all together. I now see that throughout my life I was being prepared for this project. I take no credit for the message, I am only the messenger.

GOD knows that we are flawed, yet HE loves us anyway. HE knows that our tendency is to go the other way, yet HE loves us anyway. HE accepts us the way we are because HE knows we cannot change ourselves. HE knows HE is the only one that can direct us back into harmony with the ordered system that HE created. By ourselves, we are lost, we have no 'TRUE', fixed reference.

We should not be self righteous believing that we know what is best for us. Believe me, any one of us is capable of committing any sin or find ourselves in difficult circumstances, so we should not look at others and say, 'How could they do such a thing'?. Hence the quote 'There but for the grace of GOD go I'. GOD is the only one that keeps us from being the worst we can be.

We have too little information and knowledge to be able to make some life changing decisions by ourselves. We need HIS help. We are like a small blip in the annals of time. We are insignificant in the overall order of things. HE is the only one that has the power to give us any significance so we have to look to HIM for guidance. Who better to look to for guidance than your HEAVENLY FATHER.

'GOD chooses ordinary people to do extraordinary things'.

DEFINITIONS AND CONCEPTS

H ERE ARE SOME 'definitions and concepts' to keep in mind as you
analyze the principles in this book.

Reference

The reference defines the origin as well as each subsequent point
in an ordered system. This is a critical component in any ordered
system. The reference must always be active to maintain influence or
control over the whole. It must be set by an intelligent source. This
is a universal truth.

The absolute reference governs the laws of matter as well as the
way we, as humans, behave. It is the source of the fabric of both the
mental (spiritual) and physical manifestations of the universe.

When awake, we are aware of where we are at all times. If we are
not, then we are in trouble; we are lost. We are always aware of our
location because we have stored references from our experiences to
which we have multiple access. We navigate using references both to
locate where we are and in the way we process information. This is
how critical it is to always have a reference. It is important that we
understand the significance of a fixed reference as this is an essential
component in maintaining order.

If we want to start on a journey, to go from one place to another, we must take the first step. Then we continue in a direction which we know will take us to our destination. Through all this, we need to keep in mind where we are going and use references along the way. This is a natural process. If, however, we lose our reference, we become lost. We will never be able to continue on our journey if we don't find a reference on the original path so that we can regain our bearing. This is a simple illustration but it applies to all ordered systems. We have no idea where we are until we find a reference in the ordered system that we are navigating.

If we become lost, we have to ask someone for directions. We have to communicate very clearly where we want to go and the person we ask must know the way. We can then be guided back to the original path to get to our final destination.

Without a reference, we are unable to process information. Our brain must search for and find a reference in order for us to think and interact with the environment. Intelligence allows us to perform the process of thinking and to gain and use knowledge.

Life is all about references. Without references, we would be unaware of our existence. In the absolute sense, without a reference we would not exist. GOD THE FATHER, GOD THE SON AND GOD THE HOLY SPIRIT have always existed. Since we did not always exist, for us, there must have been a beginning, a creation that started in the mind of GOD. This is our initial reference and is fixed in time. Our experiences then begin in a sequential order, fixed in time. We are then born on this earth and our experiences then begin in a sequenced order fixed in time. Memory allows us to tap into that sequence in an ordered manner.

Our brain is the medium where all of this takes place. It is designed to record this sequence and for us to access this information at will. Our brain allows us to breathe, walk, talk, hear, smell, feel, think and solve problems. The brain is also the connection between us and the material world in which we live.

Our 'BORN AGAIN' recreated spirit is the gateway to the 'SPIRITUAL REALM' where all things are possible. It connects to the 'SPIRITUAL REALM' where ideas originate to initiate our creative ability. Much of our knowledge and wisdom is communicated to us in this manner.

Our absolute reference is spiritual and is eternal. Using the absolute reference as the point from which everything originates, there is nothing new under the sun. Our universe has always been part of GOD'S plan and at some finite point in time, HE spoke it into physical existence.

Here are some everyday examples of fixed references being used to regain order and perspective in an ordered system.

Computer Reboot - program goes back to its reference point to regain accuracy.

Negative Feedback—Part of the output fed back to an input stage to reduce distortion. This cancels unwanted signals introduced by the circuit. Differences between input and output are cancelled except for proportional input amplification. The input signal is used as the reference.

Absolute Zero Temp- thermodynamics. Temperature at which all atomic movement ceases. This is the 0 energy reference.

0 - Zero - mathematics —The reference

In any experimental test, there has to be a control or reference that is the standard to determine if the experiment has succeeded or failed. Even our name and signature are types of references.

The universe has to have a singular reference otherwise there would be total chaos. It all has to be coordinated under one unifying force as this is the only way it can operate in harmony.

We have shown this by looking at isolated ordered systems and noting that each has to have a fixed, singular reference for the whole to act as one. If there are several subsystems, as in the human body, then they all have to be under one central command in order to operate in harmony. The control center must be intelligent. This feature is critical in any ordered system. In the human body, the control center is the brain.

Reference is an important component of any ordered system. However, what is also critical is that the 'TRUE' reference be used. If the TRUE reference is not the foundation of the ordered system, then it will not be in harmony with universal 'TRUTH'. This would mean that we have developed our own truth which will lead nowhere but to eventual failure.

If you don't have a reference you are lost. If you don't have the 'TRUE SPIRITUAL' reference, you are truly lost. There is only one 'TRUTH' and one 'TRUE SPIRITUAL' reference.

In 'SPIRITUAL' applications, there is only one 'TRUE' reference. In material applications, we can fix our reference depending on what we are trying to accomplish. The important thing is that once a material reference is set it can never be changed for that system. We only have the option of deciding where we want the reference to be.

When it comes to 'SPIRITUAL' references, we are not given that option. We cannot fix or change 'SPIRITUAL' references. These are fixed by GOD. This is the reason why once we become disconnected from our 'SPIRITUAL" reference, we are 'SPIRITUALLY' lost and only 'HE' can lead us back to the 'TRUE' path.

Instinct is a programmed reference and is common in living organisms. It is from the source of ABSOLUTE intelligence. If there is no fixed reference, order cannot be maintained.

Once the reference is set in a manufacturing process and the sequence is initiated, any changes must be communicated back to the original fixed reference. For any given ordered system, there is an active connection between all components and subsystems. As long as the change is in harmony with the component to which it is connected, it will be in harmony with the system as a whole. The change must therefore be of an intelligent source which is able to see the big picture and understand the relationship between all components in that system.

Spiritual Realm

It is only logical that there is another realm which is at a higher level than the material realm, as the material realm has not always existed and came into existence at some finite point in time. Also, the material realm alone does not fully explain the order we observe in our universe. In other words, the order that is evident around us cannot be fully explained by the material laws. Material laws control only matter. It would require control that is significantly more far reaching than those that govern matter alone.

Life is also another ordered process that cannot be explained by material laws alone. Material laws are designed for the stability of matter and such stability requires fixed references. However, these fixed material references are not apparent as a function of 'living matter'. Matter, in general, is subject to the law of entropy which degrades matter toward randomness. The references for life oppose entropy and must have been introduced by an intelligent source, the same source that developed the absolute laws that govern matter, as matter is designed to be used as the raw material for all living matter.

Therefore, for matter to develop life, other laws would have to be at work, neutralizing entropy or any other laws that are in conflict with the laws governing the development of life. Such laws would also have to have originated from the absolute reference that defines all things both MATERIAL and SPIRITUAL. These laws are designed to combine the 'MATERIAL' and the 'SPIRITUAL' in perfect harmony, suggesting the same intelligent source.

The MATERIAL REALM had a beginning so it is temporal. This was 'The Big Bang', as some refer to it. The initiation of the 'MATERIAL' realm must have been some external source which I refer to as the 'SPIRITUAL' realm. Even though we do not see it, it exists as there is much evidence of its existence. We have had numerous interactions with this realm which cannot be ignored.

The origin of our temporal, 'MATERIAL' world cannot be explained in any other way. It must have come from another realm, an ordered realm that pre-existed our 'MATERIAL' realm. This realm must be eternal, with its source (GOD) absolute in knowledge and intelligence. Only such a source could create a universe. This is also the source of limitless knowledge, some with which we have been blessed. We also have the ability to recognize and use this knowledge, our intelligence.

In the 'SPIRITUAL' world, absolute order is imperative. It is the realm where virgin thoughts and ideas originate. We see changes in our world that occur from day to day indicating growth in knowledge such as the vast technological growth we are currently experiencing. We, as human beings, may not have had this knowledge before, but it always existed. We are only tapping into the source. It is being revealed to us as part of GOD'S natural order of things. It is not new. It is only new to us.

We may think that it is of us but this is not the case. It always existed as TRUTH. It is eternal. It is absolute. Absolute knowledge and intelligence have always existed in the SPIRITUAL realm and is gradually revealed to us in the MATERIAL realm.

To be in harmony with the 'SPIRITUAL' realm, we need to follow 'SPIRITUAL' laws. These have been communicated to us as ethical guidelines (the commandments) in the Bible which we find impossible to fully obey without the help of JESUS and THE HOLY SPIRIT. This is because we were given free choice, we sinned, and have lost our true reference. These commandments direct us to focus on complete harmony in the way we live and interact with each other.

Eternity

Eternity 'IS'. Eternity has no beginning and no end. GOD refers to HIMSELF as 'The Alpha and the Omega'. He is eternal. This is a very difficult concept for us to grasp as we all live in a temporal world. In fact, it is impossible for us to fully understand it. We can, however, look at what factors or components must have had to be eternal for the universe to be possible.

Intelligence, as it relates to GOD, must have been eternal because time cannot explain the concept of intelligence which is not a function of time, and is of GOD. Similarly, absolute knowledge is also not a function of time. It must have always existed. The accumulation of knowledge in our world, however, as is revealed from GOD, is a function of time. The rate of knowledge gain was initially gradual but now appears to be growing exponentially.

GOD'S absolute intelligence always was and always will 'BE'. We inherited intelligence from our GOD. We only get more intelligent when our ability to process information improves as a result of changes in our brain function. This is part of the growth and maturity process. It is a function of the material world.

Intelligence is an absolute concept but the ability to accumulate and use knowledge varies between species. Man has been given a great gift, the ability to accumulate and use knowledge to a greater degree than any other specie. Our ability to gain and use knowledge is limited by brain function and this is the main variable. All the information that is available already exists and has always existed, we only discover it.

GOD has absolute intelligence which means HE is not limited by a brain in HIS ability to use knowledge. HIS knowledge is also absolute. With unlimited intelligence comes unlimited knowledge. Absolute knowledge and intelligence have always existed in GOD. They are eternal. The only change is that now we have been given the opportunity of these being shared with us from GOD at His desired fate to release.

Purpose

What do you think is your purpose in life? One of the first clues is where you were born. Also, your nationality! Whether you are male or female! What gifts were given to you! What situation you find yourself in at any given time in your life!

If this universe was created by an intelligent source, that source must have had a plan and a purpose for everything in it. It is only logical to come to that conclusion based on how an intelligent mind reasons. If we look at ourselves, we see that we do plan our lives and make an effort to have a desired outcome for our undertakings. We are made in the image of our intelligent creator, GOD.

Material

This is the realm in which we live. The realm in which our universe exists. It is made up of space and matter. We interact with it through our five senses. We can see, touch, feel, smell and hear. In this realm, the laws that govern are fixed and we cannot change them or break them.

Singular

Singular, in this context, means from a single source or the same source. Creation manifests certain patterns and similarities that can only be explained by a singular origin. All ordered systems are in harmony which is indicative of a singular source.

Matter

Matter is anything that occupies space. Even light occupies space as light rays are photons which are very tiny particles. Some may argue that light is not a particulate but a wave form, but it still occupies space. It is a form of energy.

Matter was created and creation is complete. We have not been given access to the ability to create something from nothing. We have, however, been given the raw materials to use as we see fit.

In physics, it is said that 'matter cannot be created or destroyed'. Scientists have demonstrated this theory. We cannot create matter and we cannot destroy matter. We can only change it from one form to another, as we discover in the study of thermodynamics and nuclear science. This is done at thresholds that are fixed, under given conditions. Knowing these thresholds we are able to initiate changes to get the desired results.

Using this knowledge we are able to make a nuclear device by careful preparation of the raw materials and controlling the conditions so that the thresholds are met. Our aim is to obtain a chain reaction, but the path to success is very narrow and must be closely followed.

Matter is either ordered or disordered. I will consider randomness as a form of disorder. However, disorder is more than randomness. Whereas randomness, by definition, has no bias, disorder is biased toward the negative.

We live in a material world and we know that matter cannot create itself or come from nothing. The law of physics that says 'matter cannot be created or destroyed' implies that, if you have a void, matter cannot suddenly appear in that void. Any matter that exists in our world must have always been in existence in one form or another, since its creation.

We may therefore conclude that if there is a void and matter suddenly appears in it, then it must have come from somewhere else and, if created, it must be of a source outside that void.

Matter cannot create itself. If it could, we would see evidence of such creation in this realm. The ability to create must therefore beof another realm. Creation meaning something from nothing.

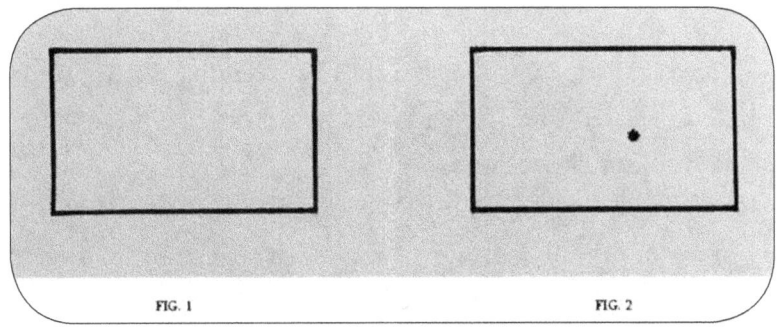

FIG. 1   FIG. 2

Let us look at the diagrams above to try to explain this concept.

In the first diagram, Fig. 1, there is nothing, which represents a void. There is no solid, liquid, gas or energy present. No matter present. The second diagram, Fig. 2, has a 'dot' representing some form of matter. It does not matter what it is but, if it suddenly appears it must have come from somewhere. Now, if before there was only a void, then the dot must have come from somewhere else. Its source must have been something outside that void. In other words, an external source. This is the only way it could have appeared in what was once a void.

We can look at our universe in the same way but on a much larger scale. Its source must have been outside our universe.

Properties

Properties are characteristics that are unique to a particular material, process or system. They are sometimes used to define the material or process because of their unique characteristics.

If we look at 'salt', as an example, some of the properties would be that it is white, it is crystalline/granular, it is a solid at normal temperature and its taste is 'salty'.

Something

If we look at matter and the concept of something versus nothing, we have to start with a reference. There can only be 'something' if there is 'nothing' or vice versa.

There has to be 'something' to compare it with in its own category or realm otherwise it is meaningless. It would not be relevant. (something refers to matter as we have defined it)

'Nothing' and 'something' are at the opposite ends of the spectrum in our realm. They are mutually exclusive. But, one cannot be perceived without the other being taken into consideration. We are unable to appreciate 'something' unless there is 'nothing' to compare it with. This demonstrates the importance of reference. We may look at space as representing nothing and matter representing something. Similarly, we would not understand light unless there is darkness.

If we say, 'in the beginning, there was nothing", then we have defined nothing, which will become our reference. But, how can we perceive nothing unless there is something.

If there was 'nothing' in the beginning, a new realm must have been created when 'something' was created. The 'something' being matter as we know it today. The concept 'nothing' was defined the instant 'something' was created or vice versa. To define 'something' the concept of 'nothing' had to be invented. This could only have been perceived by a source outside the material realm as there was nothing in our realm to perceive it.

In the beginning, there must have been some other realm in order for 'nothing' to have been perceived. Remember, 'nothing' only became relevant when 'something' was created.

The preexisting realm, I believe, is the 'SPIRITUAL' realm. The realm in which we now live is a new realm, the 'MATERIAL' realm. That initiating source had to have been outside the realm of 'something/nothing' as we know it today.

Now, if we look at this concept in the absolute, SOMETHING must have existed in the realm from which 'something' in our realm was created. This 'SOMETHING' is a different 'something' as we define it in our realm, the material realm. It must have been eternal since, if it had a beginning, there would have to have been something external to initiate it. This leads us to conclude that this

'SOMETHING' must have always existed- the SPIRITUAL realm. This is the only explanation.

In conclusion, the 'SOMETHING' from which our universe was created must have always existed since only such a source could be the absolute initiator. We have no understanding of what eternal means but our interpretation is that it has no beginning and no end. The "Alpha and the Omega". (This is how GOD describes himself)

It is impossible for us to perceive 'eternal' as we are constrained by our experiences which are of a temporal world. But 'SOMETHING' must have existed prior to the creation of our universe otherwise our universe could not have come into being. If we look at the product of this prior existence, our universe, an ordered system, then there must have been INTELLIGENCE to define order in that realm. That order was then commanded to the 'material' realm to manifest itself as the order we see in our universe. This is as far back as I can rationally go as that realm is beyond our sense of reasoning.

Nothing

Nothing, as we define it in our world, is a 'void' where no matter exists. Like in outer 'space' where there is no matter. Here, space is also loosely defined as, even where there may be no visible matter, the 'space' is occupied by small particles separated by large areas of void. What we are actually doing is comparing it with what we know as 'something'. But 'something' cannot come from a void. To initiate 'something' in a void there must have been an outside source, independent of the initial void. 'Nothing' became relevant or valid when 'something' was created. In other words, in our world, both 'something' and 'nothing' are dependent on each other for validation. One cannot exist without the other.

We can only imagine what our brain allows us to perceive. Since we are a part of the 'something' it is not possible for us to objectively examine this (something/nothing) concept and come to any meaningful conclusion. We would need to be outside the system to think objectively. Unfortunately we will never have that opportunity in this world.

Positive

In our universe, we see evidence of attraction and repulsion relative to the very building blocks of matter. An atom consists essentially of protons, neutrons and electrons. Protons are positively charged and electrons negatively charged. Neutrons have no charge and are neutral, as the name implies. Neutrons are believed to bind the all positive nucleus of the atom/molecule together despite the all positive charges on the protons repelling each other. However, as the nucleus gets bigger, it becomes more unstable and eventually so unstable that it breaks down at a certain threshold, as we see in nuclear isotopes.

The positive and negative charges on the protons and electrons, respectively, are the forces that bind the electrons to the all positive nucleus (protons) and also make it possible for new compounds to be formed in chemical reactions.

The movement of free electrons in a conductor manifests itself as an electrical current. Electrons move towards the positive pole since opposite charges attract.

We also use the term positive to describe something good and negative to describe something bad.

Negative

Negative is only a way of expressing the converse of positive. We have found that negatively charged electrons are attracted to positively charged protons which means that, relative to the electron, the proton has an opposite or positive charge. The reference may be either the proton or electron but once we set this reference it has to be fixed for order to be maintained. Electrons are in one category and protons in the opposite category.

Protons, neutrons and electrons are the main building blocks of matter and combine in fixed proportions to form different molecules and compounds. It should be noted that the combination of atoms (protons neutrons and electrons), exist sequentially in the periodic table. Compounds are formed from atoms combining, but only relative to their valencies. Both the sequencing of the periodic table and valency manifest order and consistency.

Good

Good is anything in harmony with the universal order, both material and spiritual.

Bad

Bad is anything in disharmony with the universal order, both material and spiritual.

Truth

There is only one TRUTH. TRUTH is FACT as it relates to our very existence and interaction with each other and the universe. There is no alternate truth; there are no alternate facts. Man is a singular system with multiple subsystems, all connected. For us to live in harmony, with the ability to communicate with each other, there can only be a single source controlling it all. That source defines and executes 'TRUTH'.

In our judicial system, we seek to find the truth before making a judgement. We may not always find it, but we should always do everything to uncover the truth. This is instinctual because we are products of order. However, if our reference is not (absolute) TRUE (THE TRUTH) we can never discover 'TRUTH'. The only 'TRUTH' originates from our source (GOD). Any variation from this is false. We must therefore only seek 'TRUTH' from our source.

But, how do we find the source of TRUTH? We do have access to this source (GOD) but unfortunately most of us do not know the source and are not able to identify HIM. We need to faithfully ask the source that the TRUTH be revealed to us. This is what faith is all about. GOD is the source of 'TRUTH'.

This is where our creative abilities originate. Is it not logical that, if we get our intelligence from our source, the source would have more intelligence than we do? If so, wouldn't the source have made a way to communicate with us and vice versa?

As I mentioned before, there is only one TRUTH and only one TRUE source. This is how order is created, from TRUTH. Absolute TRUTH. Absolute or universal TRUTH defines all perfectly ordered systems.

Today we live in a world where 'Fake' news is a misleading influence on society. There is also nothing such as 'alternate facts' to which we sometimes hear people refer. The TRUTH is absolute and there are no variations on the TRUTH. Facts are facts and there are no 'alternate facts'. We are sometimes unable to get to the truth but the failure is on our part. For any given situation the truth never changes.

Fabric

This is a metaphor indicating that the ordered systems of the universe are interwoven. The fabric extends from a fixed point into infinity in all directions.

Intelligence

Intelligence is the most important component for the creation of an ordered system. It is the initial requirement for developing an ordered system. Intelligence is the ability to create order. Intelligence defines order.

In any ordered system, the designer must have some concept of the finished product and its intended function. This takes intelligence. Also built into the design of a system is the ability to adapt to changing environmental conditions, to ensure that the system will continue to operate as designed.

Humans use a similar principle when we design an ordered system. We anticipate changing conditions and compensate for those in the design. In order to do this, we build into the system the ability to recognize the perceived environmental changes and then to initiate the steps to compensate, so that the system will continue to operate as designed, in a stable state.

The design of living organisms indicates that adaptive changes were anticipated by the designer and built into the initial design of the system. Any changes that occur, whether adaptive or 'evolutionary', were done by the intelligent designer.

In our case, when we design something, we sometimes do not anticipate these environmental changes and so need to modify the initial design based on failure experience. Failure may also be the result of flaws in our initial design. We must then modify the design to correct the design flaw. All this takes intelligence which is the most important requirement for the design and development of an ordered system.

If we assume that our universe is an ordered system, then it must have been created by an intelligent designer.

Intelligence allows us to understand the concept of sequencing and timing thus giving us the ability to recognize disorder and to create order.

Intelligence applies in every discipline and every aspect of our lives.

We have to comply with the laws and rules of the ordered system in which we find ourselves, otherwise we would be out of harmony and unable to integrate. The laws governing matter are fixed and never change. All we can do is discover and familiarize ourselves with them and use them to our advantage. This is the process of gaining knowledge. Intelligence gives us the ability to understand these laws which we then use when we create and invent, using matter which is subject to them.

The intelligent designer must develop a plan and execute it from beginning to end. To do so, a reference must first be set and this reference must be fixed. Execution of the plan is done using continual feedback to ensure that the plan is being executed in the right sequence, with monitoring for errors at every stage. The intelligent designer must be a part of the process in order to know when the plan is complete and whether any other changes or modifications are necessary.

Intelligence, by definition, is order creating, it is logical, it is rational, it is systematic. It is an ordered system. Only order can create order. Only intelligence can create order. Intelligence is able to recognize disorder and also has the ability to change it into order. But, there must first be a fixed reference with subsequent sequential, positive steps and continual feedback to ensure that progress stays on track. Feedback is built into the system.

This is what we refer to as checks and balances or quality control. Intelligence gives us this ability.

We use the thought process that we have inherited from our creator. We think rationally and logically to solve problems and to act. We develop a plan and then execute it in a logical sequence. We inherited this method as the natural way to approach systems design. Wouldn't it then be logical to conclude that this is how nature works. You see, we are a product of nature. Then, evolution, which is random, does not fit this profile. The process of evolution would have to have intelligence but there is no evidence that there is feedback or planning, only chance change and adaptation resulting in random or natural selection.

Based on the way we think and reason, there is a plan with a preplanned outcome. This is far from random.

If one thing is changed in an ordered system it affects everything in which it is in immediate contact and sometimes other areas as well. In order to compensate for this change, modifications will need to be made in the affected parts of the system. It takes intelligence to recognize and make these changes as they will sometimes be significant. In the case of a living system, the changes will also need to be communicated to subsequent reproductions (in the DNA). This process is very complex and unlikely to be random in nature. It would take a source with complete knowledge of the system design, awareness of the change and ability to modify the design to compensate to the degree necessary.

We see intelligence in our world but there is no indication that any element on earth contributed to the creation of this intelligence. Matter has to be instructed on how to be ordered. I think it is reasonable to conclude that the intelligence we have inherited is not of this earth as there is no evidence of such intelligence in the basic building blocks of nature. If it is not of this earth, then its source must be of something external.

Using our intelligence, we create and invent things that never before existed on this planet. It must therefore have been initiated by the same external source which directs us to study our environment where we will find the raw material that will make things that are imagined become real. It tells us that all we have to do is search and we will find it. If we can imagine it, we can make it real. This is creativity in its virgin form.

INTELLIGENCE does not evolve. INTELLIGENCE exists. It 'IS'. It is absolute. GOD is the source of all intelligence. However, as far as it relates to us, our individual abilities do vary. Each of us is a medium with the ability to be intelligent, hence the ability to gain and use knowledge. Our species has the greatest ability to be intelligent based on its design. It was made in the image of GOD.

Within our species, intelligence is distributed in the form of a normal distribution. Most people are of average intelligence, then there are those at the extremes. The normal distribution represents an ordered system, with the distribution of the sample or group being about equal on both sides of the median (reference). This is the reason why we have named this statistical measure of distribution 'normal' because this is how specific features of groups manifest themselves in nature. None in any group is exactly the same and specific features in that group can be represented by a normal distribution.

(See diagram of normal distribution on page 81)

If you look at the curve of a normal distribution, you see that it is a combination of an exponential increase and an exponential decrease on either side of the reference (median) or statistical norm. Neither side reaches infinity as in the typical exponential graph. They flatten and meet at the maximum then decrease back down to zero.

Intelligence can only be derived from something already intelligent. GOD, the source of all intelligence is eternal. Absolute intelligence does not evolve. It already exists in the infinite.

Artificial Intelligence

We are now in the age of artificial intelligence. Our reference is the human brain or mind and how it controls the body and solves problems. We are trying to simulate how the human brain works with its subtle abilities to display logical and rational reasoning in problem solving and make the best decision in any given situation.

To be comparable with our brain, artificial intelligence must be able to switch command and control from one situation to another, as it is confronted with changing issues, making split second decisions where needed. This is a difficult prospect. Currently, artificial intelligence can be designed to do a single task very well but does not have the flexibility of switching to another task, using a process of logical reasoning, for which it is not designed. It therefore lacks the flexibility of the human brain in these circumstances.

One of the reasons for this shortcoming in artificial intelligence is that our brain gets input from five sources, our five senses. The brain is able to prioritize these inputs and make split second decisions as to what it needs to react to as well as how to react. What we should do first, followed by the next sequential step and so on, until the response is complete. There are also responses at the subconscious level from other inputs. All this is done by one small mass of cells that has complete control over our body functions.

We can identify danger by vision, sound, scent, touch and taste. We are continually monitoring all these inputs and are always ready to react to any situation presented to us. We immediately know how to prioritize in any situation with which we are presented. Sometimes our reactions are subconscious in cases where rational thought process is not fast enough.

This is a challenge for artificial intelligence. But this is not the only obstacle. We have complex emotions such as love and hate and empathy. These are very difficult to quantify and hence to program into artificial intelligence.

Order

How we recognize order is fairly evident. We look for consistent patterns, cycles, harmony, theme, balance, aesthetics, timing and sequencing of events.

We are programmed with the ability to recognize order using the above clues as a guide. References define who we are and provide evidence as to our source. They tell us that our source is ordered. They reveal the character and essence of our source.

Order is the opposite of randomness or disorder. Order includes anything we do that involves intelligence and knowledge and utilizes a logical sequential process based on a fixed reference. This would be applicable to the design and fabrication of something as simple as an arrow head to the complex manufacturing of an automobile or a spacecraft. The only difference is that the arrow head is significantly less complex than the automobile and the automobile significantly less complex than the spacecraft. They each require different levels of intelligence and knowledge.

We must start from somewhere, but first, we have to have intelligence so that we are able to obtain, retain and use the applicable knowledge. Intelligence is therefore a critical requirement. It is the first and most important requirement in the development of an ordered system.

We were given intelligence with the ability to problem solve. We should always remember we all came into this world and we contributed nothing to the process. We were given intelligence, so intelligence must have existed before us.

We have everything we need to survive and grow. We are able to connect with our environment through all our senses. This is as complete a set of senses as we could desire. So, did these senses develop randomly as suggested by evolution or was it an ordered-development? Could their development have been stimulated by the environment? That seems irrational as the process of evolution would not know how or where to begin.

Order is apparent both in material and spiritual systems. In fact, the material order is modeled after the spiritual from which it originated. Material order is a reflection of spiritual order. They were both developed with strict laws that govern the behavior of the systems that are subject to them.

An ordered system is interconnected. The interconnection is active. It has to be! Every element in an ordered system is connected to the other either directly or indirectly. This is how order is maintained. This is critical in both the 'MATERIAL' and the 'SPIRITUAL' worlds.

The reason why there is no absolute SPIRITUAL order in our world is because we are disconnected from GOD, our 'SPIRITUAL' source, our 'SPIRITUAL' reference.

GOD had to disconnect us because the 'SPIRITUAL' world has to be perfectly ordered and uncontaminated. Once we sinned we had to be disconnected.

Order and intelligence are synonymous. Order is what intelligence is all about. It is true that some intelligent sources set out to create disorder, but this is intentional. Such disorder is with a negative bias of which the source may be fully aware and consciously pursuing.

Sequence And Timing

To create order, in our world, requires sequencing and timing. To monitor sequencing and timing, there has to be an intelligent source in both the design and execution processes.

Disorder

Disorder may be similarly defined as order but with a negative bias, and also of an intelligent source. Disorder implies it is with the specific intent to disrupt and negatively impact an ordered system. Disorder may also be described as the effect of a random force on an ordered system. Disorder is the result of a negative influence on an ordered system.

Our source, GOD in heaven, is where perfect order exists. Disorder entered into our world because of original SIN.

Randomness

Randomness implies disorder but with no particular bias. There is no bias because there is no fixed reference. There is no single force influencing the final outcome that significantly outweighs the others involved.

Randomness is inert unless acted on by an intelligent source. The intelligent source extracts order from randomness by giving it a fixed reference. It now becomes part of an ordered system and under its control. It now has direction and purpose.

We recognize that something is random because we cannot comprehend it. Randomness is foreign to an ordered mind.

Knowledge

We gain knowledge by observing, understanding and recording the patterns and laws we see in nature and the universe. We cannot change the laws that govern them. These are fixed. However, by understanding them we are able to use them to invent or create ordered systems that we use to enhance our quality of life. Our brain has the ability to identify and store these patterns or bits of information in the form of knowledge, for later use.

The order we see in nature and the universe existed before we came into being. These consistencies that we refer to as laws suggest that there was an intelligence that created them.

All the knowledge that we now have and will discover in the future, has always existed. All we are doing is discovering and learning about our realm. All scientific discoveries are only being revealed to us now but these truths have always existed and have always been known.

Entropy

Ordered matter always gradually declines into disorder or randomness and loss of energy if not interrupted by an instruction for order to be initiated or maintained. This gradual decline is known as entropy. The instruction for order has to be of an intelligent source, a typical example being the manufacturing process. Entropy will continue until a stable state is reached and, after this, random change will continue to occur. Entropy is the second law of thermodynamics.

Error

Error is disorder. Error multiplies! If an error is made at the initial phase of an ordered process, it is multiplied in subsequent stages and therefore amplified as the process progresses. This is why it is critical for quality control procedures to be in place, at every stage, from start to finish in any manufacturing process.

In a manufacturing process, at each stage, the quality of the individual components must be checked, before assembly begins, to ensure they are within the design tolerance. If a component is found to be out of tolerance, it is rejected. All components in the assembly must be within the specified tolerance otherwise the finished product will not be acceptable.

In other words, it takes very close monitoring at all the production stages to make an acceptable finished product. Error creates disorder and disorder multiplies, if left unchecked.

One reason why error multiplies is that the reference is always changing. For example, if we are making a batch of similar components, after the first error, if the original reference is not used, error will begin to accumulate. As manufacturing progresses, if a new reference is continually used, as opposed to the original reference, errors will continue to increase.

Using the original reference in all cases will ensure minimum error. It is also important that the reference be a true reference and be accurate.

If a story is told by one person and subsequently by others, each time the story is told there will be variations on the true narrative. These variations will increase as the number of people telling the story increases. By the time it gets to the 10th person, it will not be the same story. The reason is that the reference is always changing and so error increases each time the story is told. The references become the different people telling the story.

Aesthetics/beauty

Aesthetics play an important part in our culture. We appear to have an innate sense to appreciate beauty that sometimes transcends cultural boundaries. We inherently like things we consider beautiful and dislike what we consider ugly. Here again, there must be some universal reference with which we compare in order to come to a conclusion as to what we consider ugly or beautiful. Some of the things that probably guide our sense of judgement are symmetry (balance) and color, that appeal to our emotions in some positive or negative way based on our individual taste.

We associate beauty with order and ugliness with disorder. Beauty has a positive emotional effect on us. It affects our state of mind and hopefully evokes something 'good'.

CONCLUSION

THE EXAMPLES I'VE given in this book are only a small sample of the ordered systems that we encounter in any discipline that we may analyze. This indicates that all ordered systems must have these basic components (intelligence and a fixed reference), and they define such systems. To test this theory, you should select any ordered system that you encounter and analyze it to see if you can verify the theory. You will find that, on analysis, any ordered system that you choose will have these basic components, and there is no exception.

I wrote this book to convey a message that answers some of the questions we all have about our existence. The included definitions and concepts should help give you a better understanding of the principles that govern our material and spiritual worlds. Both manifest one common element, and this is order. This order must be defined by intelligence, as this is the only attribute by which order or disorder can be recognized.

In our world, we are always trying to create order and is the desired end product. This is reflected in all aspects of our lives and is something we are always striving to achieve. As indicated in the definition of intelligence, this is the most important component of any ordered system. Intelligence defines order and so must be the source of any ordered system.

Since the universe manifests itself as an ordered system, it follows that it must be of an intelligent source. It is also clear that creation is from a singular source, which scientists refer to as the Big Bang. Since the Big Bang produced an ordered system (our universe), this must have been the product of an ordered source, and by definition, that source must have been intelligent. This is because order cannot be produced from disorder or randomness unless directed by intelligence.

Finally, we can conclude that the source that produced the universe must have been intelligent because it produced an ordered system. This concept is at the core of what I am trying to convey to you, the reader. Based on how God describes Himself, He is the only one that fits this profile and thus, puts everything in perspective.

AUTHOR'S BIO

I WAS BORN IN Oracabessa, a small town on the north coast of Jamaica. For those who are familiar with the popular vacation spots on the island, it is thirteen miles east of Ocho Rios.

I have always had several hobbies and would switch from one to the other depending on my mood at any point in time. My hobbies included drawing, from an early age, which led to oil painting. I also made model cars, boats and planes as a teenager. I then became interested in electronics and made amplifiers and preamplifier, from circuits printed in Popular Electronics, to enhance the fidelity of the music of the time, as music had now become one of my interests. Woodworking was my later interest and I made furniture and speaker enclosures.

I enjoy using my hands and would sometimes skip meals from being totally absorbed in a project. This gave me an understanding of the properties of materials and the knowledge required to plan, initiate and complete a project.

This experience influenced me in my choice of career to become an engineer. It seemed natural as I was always interested in how things work and tried to gain the level of knowledge required to successfully complete my projects. This later led me to seriously consider what initiated the universe and the forces needed to maintain order. It is apparent that the universe is ordered, but how did it get that order?

This book is designed to give the reader a rational explanation as to the origin of the universe, from a scientific viewpoint, based on its physical manifestations with which all of us are familiar. We will see that the same principle is also true for spiritual order.

REFERENCES

The Holy Bible.

G OOGLE TO VERIFY the equation of the formula for the gravitational force on a body and the theories associated with evolution.